TO MY PARENTS

PREFACE

I first became acquainted with the Harry Orchard story as a young student in a high school history class. It became a life-long interest, culminating in a five-year research period that started in Boise, Idaho, with court records, manuscripts, and file material. Then I went on to the great silver mining district in northern Idaho, the copper producing community of Butte, Montana, and finally to the gold producing camps of Cripple Creek, Colorado.

In the interests of historical accuracy, I left uncorrected errors in grammar, punctuation, and spelling. Conversation was kept in its original form as recorded, and I have not knowingly presented events or facts without some basis in fact. In many instances, the truth has proved to be stranger and much more fascinating than fiction.

Perhaps the entire story of Harry Orchard will never be known, but the conclusions of my research have indicated that events happened much in the manner the professional assassin stated they happened in his confession, documents, and other writings. The sheer enjoyment of research was coming in contact with the many people who assisted in so many ways. Former prison wardens, John W. Snook, Ira Taylor, and L.E. Clapp, were most helpful, as were Henry L. Day, Mrs. Francis Paulsen Moore, Mrs. Harry Marsh, and Idaho Attorney General Tony Park. A fitting climax to it all was the opportunity to examine the Haywood trial evidence, still intact through the years.

Much generous assistance was rendered by Dr. Merle Wells, Director of the Idaho State Historical Society, and Mrs. Alys Freeze, Head of the Western History Collection of the Denver Public Library. Valuable assistance was also obtained from staff members of the Utah State Historical Society, the Eastern Washington State Historical Society, the State Historical Society of Colorado, and the Nevada State Historical Society. Librarians and assistants of the Spokane Public Library, the Idaho Falls Public Library, the University of Idaho Library, and the University of Montana Library, also gave freely of their time, material, and facilities.

A special thanks must be extended to Mrs. Marilla G.

Grimmett, Mrs. Mildred Hoggan, and Mrs. Myrte Armstrong, for advising, proofreading, and correcting the manuscript, and to my wife Peggy, Mrs. Ann Miller, and Mrs. Nancy Brizzee who tirelessly completed the necessary typing and retyping.

INTRODUCTION

In the winter of 1906, a man charged with the murder of a former state governor dictated a confession of his life in crime, thereby completely astounding a nation. The story itself seemed unreal, and the magnitude and character of it were not even comprehensible to the majority of people who could not or would not believe the circumstances and the events described could actually exist.

Few informers in the Twentieth Century have created more impact by their confessions. It was a forerunner to the later exposures of Abe (Kid Twist) Reles in the operations of Murder, Inc., and of Joseph Valachi who opened up the secrets of the Cosa Nostra.

Harry Orchard admitted to being a multiple murderer, arsonist, dynamiter, thief, wife deserter, liar, bigamist, and would-be kidnapper. He not only confessed to eighteen murders, but also to the attempted assassination of Colorado's chief executive, two members of the state supreme court, the adjutant general, and various other prominent men. Orchard claimed he had committed most of his crimes at the direction of leaders for the Western Federation of Miners, a labor union known for its extreme radicalism and violence.

As the nation gasped in disbelief, attention was immediately focused on the capital city of Boise, Idaho, where the confessed criminal was being held. Was this monstrous story of organized death simply the hallucinations of a pathological liar, or was it the act of a man cleansing his conscience and soul with this "death-bed confession"?

The confession resulted in the arrest in Denver, Colorado, of important union officials who were spirited from that city amidst considerable secrecy and transported to Idaho to stand trial, charged with being conspirators to the act of murder. At once complaints from the labor world echoed from coast to coast, carrying the cry of kidnapping and the deprivation of legal and constitutional rights. When the union leaders finally came to trial, it was more than merely a case of homicide. The charge, in effect, indicted an entire union organization.

The crux of the trial was centered around which part of the

confession was actual fact and which part was merely the figment of a man's imagination. The prosecution operated on the belief that the entire confession was the unmitigated truth, while the defense naturally argued that it was a string of lies from beginning to end.

The prosecution, therefore, had the very necessary legal task of supporting and proving the validity of the confession within the strict confines of state law to the satisfaction of the twelve men who sat in judgment. The state also had to prove not only that Orchard had committed the crimes he acknowledged, but also that they had been perpetrated at the specific direction of the indicted union leaders. The state could show he was acquainted with and frequently visited the union officials, but it was more important to establish the nature and substance of those meetings. Irrefutable evidence then had to be presented to the jury to prove that Harry Orchard had indeed committed the crimes he claimed and that the other accused men had directed him in the commission of these crimes.

Prior to the actual trials, the state was on a fairly sound foundation — the prosecution had arrested another self-confessed killer, Steve Adams, a partner-in-crime of Harry Orchard. His confession paralleled Orchard's and supported it on major points. The two men claimed to have assisted each other in several killings, which they both insisted were done on orders and under the direction of union officials. Adams, however, repudiated his confession before the trials, leaving only Orchard's statement to prove.

Left with one main witness, the state's position naturally was not so formidable. Now, other supporting evidence became imperative. The nation's most prominent detective agency became the chief intelligence agency of the state in its collection of information and its attempt to substantiate the state's charges.

The trials reached national and international prominence during the hot summer months of 1907. The first union leader to be tried was William D. Haywood, secretary-treasurer of the Western Federation of Miners. The chief witness for the State of Idaho was, of course, Harry Orchard. The truthfulness of his confession is still being debated after all these years. After he had taken the witness stand, one newspaper said of his testimony,

"He spoke of killing people as butchers might speak of slaughtering a lamb."

Were the accused labor officials as free from sin as they claimed? An interesting aspect of the case is that not one of the accused persons in the Haywood trial ever changed his story or admitted to anything other than what he professed to be the truth at that time.

The destiny of those involved in the case is equally as interesting. Big Bill Haywood became involved in even more controversial union activities before he died in Russia and was "laid to rest" inside the walls of the Kremlin. William E. Borah, a state prosecutor, John Nugent, a defense attorney, and Frank Gooding, governor of the state of Idaho at the time, all subsequently became United States senators. James H. Hawley, chief prosecutor for the state, was elected governor. Clarence Darrow, the defense counsel, became recognized as one of the outstanding criminal attorneys in legal history. And the man who provoked the case, the depraved mass murderer, became a prime example of how the power of religion can influence life, proving in time that he could be a respected and constructive person of the prison community. This is the story of that man, Harry Orchard.

Fig. 1. Harry Orchard, at the time of the trial.

CHAPTER I.

Assassination was a trade and murder a means of livelihood
. . . Edmund F. Richardson

The hall outside the courtroom was packed with struggling, perspiring men and women fighting for a place near the door. At fifteen minutes past one o'clock when the doors were finally opened, three burly guards were required to keep the crowd from tearing the large doors from their hinges as they surged into the room. Hundreds of people were turned away for lack of space, and a large number of them were women.

Once in session, the anxious and curious realized that the moment they had waited for was fast approaching. They stared at the two large doors of the room. Then the silence of the court room was suddenly broken by the deep bass voice of the state prosecutor:

"Call Harry Orchard."

There was a stir in the courtroom. Suddenly the doors opened and deputy Ras Beemer entered, followed by the prisoner, Harry Orchard. Orchard looked up with some hesitancy as he searched for the clerk of the court. Spotting him, he walked slowly forward. As he approached the elevated witness stand, the prisoner seemed, momentarily, to lose some of his physical control. He reached for the arm of the wooden chair, and his mottled face revealed great nervousness as he stood with uplifted hand while the clerk administered the oath.

Haywood's one eye fixed on Orchard with a look of intense hatred the very instant he entered the room. He caught Orchard's eyes, and for several seconds the two stared at each other. Finally, Haywood looked away, as if lost in thought. Meanwhile, the gaping-mouthed spectators took in every moment of the courtroom drama as they strained forward in their seats to watch the witness.

"Where do you reside?", asked the large, heavily-built prosecuting attorney who began the examination at once. As if taken by surprise, in an almost inaudible tone, Orchard stated that he was confined in the state penitentiary. Then, as his wavering voice recovered, he began to relate the bloody details of his many crimes in low, confidential tones. His manner was easy, but his steel-grey eyes took on an unmistakable brilliance.

Thus, Harry Orchard began to narrate one of the strangest and most fantastic confessions ever recorded:

> I was born in Northumberland County, Ontario, Canada, on the 18th of March, 1866. My real name is Albert E. Horsley. My father was born of English parents and my mother of Irish.[1]

Northumberland County in the province of Ontario borders Lake Ontario in that part of Canada referred to as Canada's Sun Parlor. Orchard was born near the small town of Brighton, some one hundred miles east of Toronto. Winds from the Great Lakes give that portion of Ontario a comparatively mild climate, and the fertile grey-brown soil of the region produces a variety of crops, including grains and vegetables. Beef and dairy cattle graze on fenced grasslands.

Orchard was the second oldest of eight children. They were conscientious and capable workers; they had to be in order to make ends meet. When Orchard was about ten years of age, his father bought a small farm which became the cornerstone of the Horsleys' existence. Here the family made use of all their resources, shearing wool from their own sheep, weaving it into cloth, then making clothing from it. They raised their own meat and hand churned the weekly batch of cheese and butter. The grain they raised was hauled to the local mill, ground into flour, and then hauled back to the home where it was used for cooking.

Reflecting on his early life, Orchard continually dwelt on his love and admiration for his mother. He feared his father, even though he acknowledged him as a hard worker. Ruling with an iron hand, the older Horsley exhibited little affection for his children and was not cut out to be a family man or a financial manager. The children grew to distrust the father's abilities, and the only bond that held the family together was the loving mother. Orchard was to remark, half a century later, "She was the light and the sunshine of our family."

Religious activity in the Horsley family was directed and

guided by the firm but gentle mother who had become a Quaker. Each morning as the children sat down to breakfast, she quietly read from the family Bible. The children, with heads bowed, heard the wisdom from another age as their adored mother read. Ironically, scores of years later within the prison walls, Orchard would repeat this daily ceremony just as his mother had done. Many times this murderer, when speaking of his mother, would refer to her religious attitude "as a childlike faith in God." And his eyes, old and tired, would moisten as he spoke of her.

After his confession, this professional assassin stated that he was grateful his mother had not lived to see him in his later years. In confinement, Orchard adopted the religious practices and teachings he had been exposed to as a youth. Confinement, however, in a penal institution may well have given him no other choice. Ironically, though, in many ways he finally became the man and individual his mother would have admired. Then, too, his actions could also have been the expiating of a long, deeply-submerged guilt complex.

There seem to be no other instances of criminality on either side of Orchard's family, although both his grandfathers had drinking problems. The only case of any mental disorders was an uncle who was eventually placed in an institution.[2]

Harry Orchard and his father were invariably in bitter conflict with each other. The elder Horsley, never providing adequately for his family, nevertheless exerted a strong influence over his "clan." If the kindness of human nature was found in his mother, the other extreme existed in his father. As a result of her understanding, she became a consultant and a confidante to the children.

Orchard was allowed little formal schooling, probably the equivalent of a third grade education, but this was not unusual in the 1870s. He was forced to drop out of school to help support his family. In addition to helping operate the family farm, he was also hired out to work for neighbors. His parents, also not unusual for their times, received the financial benefits of his labors until he was more than twenty years of age. However, there was constant emotional and physical conflict between the growing boy and his father.

Resentment flared as the hard years of adolescence passed. For Harry, life consisted of a series of agricultural pursuits, always a hired hand, always working for the man just down the road. Orchard strained to break away, to be his own kind of man. Meanwhile, Orchard spent many of his evenings playing cards, a pastime for which he developed a lasting passion. He soon became known as a winner, and the game of chance and the gambling spirit became a part of him.

Eventually, he selected a pretty Scottish girl named Florence Fraser. She, by nature religious and industrious, was by trade and training a highly skilled cheese-maker. The couple wasted little time in courtship, and plans for the future were carefully drawn, a future that seemed uncomplicated and full of promise. Engaged and realizing the need for additional money, Orchard spent the winter of 1888 in a logging camp north of Saginaw, Michigan.

Often he drank until he was sick enough to believe he would die, and frequently he wished he had. Each time he vowed never to touch the stuff again, and he kept his vow faithfully until the next time around. He also tried professional gambling, thoroughly enjoying the thrill of betting, winning, and bucking the odds in drawing to a straight flush, but most of the time he lost more than he won. In meeting the town harlots, he found he could purchase a sex life as easily as he could a bottle of liquor. He was completely captivated with this irresponsible life.

Stories of unlimited opportunities that awaited the adventurous in the country still farther west held him spellbound. A man could make better money, perhaps a fortune. The kind of life he wanted was his for the asking. When he wrote Florence about the advantages and opportunities of the West, she threatened to break the engagement if he continued to entertain such ideas. And so the couple married soon after Orchard returned to Northumberland County in June, 1888, and used part of their savings to furnish the small house in which they lived.

Florence introduced her husband to the cheese-making business, and he soon realized that a person could do well in that occupation. The following year the couple rented a small factory that was badly in need of repairs. Securing a loan, they started working it. The cheese business was highly competitive in that particular part of Canada, but Florence was not only skilled, but also devoted to turning out a first-class product which soon put the business on a paying basis. The cheese factory flourished so well that by the following year they were able to buy the business outright.

Unfortunately, the experiences of Harry Orchard's youth had produced a man who could not cope with success or prosperity. This insecurity resulted in a life-long practice of grasping for undefinable limits beyond his reach and totally unrealistic. In the cheese business, Florence guided him by assuming the technical work herself and by delegating the buying and selling end of it to her husband. Once he reached a certain plateau of success, Orchard began drinking heavily and carousing. Such episodes were to occur again under similar circumstances throughout his life.

When Orchard reached the point at which his spending exceeded his income, he became desperate. After his bank credit was depleted, he juggled the company's account books and short-weighed incoming milk, buying a poorer grade of milk which lessened the quality of cheese produced. After several years of strain, the business folded and was sold to meet financial obligations.

Not even the able Florence could control the situation. In response to her many unanswered questions, he shrugged his shoulders. When pushed farther, he retorted, "Mind your own damn business," and ended the argument by getting drunk. "It was good while it lasted," Orchard reminisced in later years.

The spring of 1892 found the couple working as common laborers in another cheese factory at Wooler, doing much of the same type of work they had once done for themselves. There was more work, less money, longer hours, but it seemed to be just the tonic for their problems — particularly for Harry. He had to curtail his outside activities and even showed signs of being repentant. Once again he became a steady and a regular church member, accompanying his wife to services in the local Methodist church. There he was placed on probation but apparently was never admitted into full membership. although he did teach a Sunday school class for a short time. They settled their debts and started saving. Tension between them relaxed and sparks of their former relationship caught fire.

Then Harry Orchard was once again attracted to other women, the cardboards felt inviting to the touch as he dealt, and the whiskey tasted good. His pre-occupation with "fast living" became evident to everyone in the small town. Public opinion caught up with the dapper cheesemaker and ostracized him, and in the end his actions cost him his job.

Relations between Harry and Florence once more became strained. Short fights became long fights. Preliminary bouts became main events. Repulsive bitter words passed between them, leaving scars that refused to heal. The more Florence pleaded and threatened, the more her husband felt the need to be elsewhere.

At this time another of Orchard's interesting qualities emerged: the art of making and using friends. Throughout his life he had the ability to make friends, and some of these friendships survived court trials, long years in prison, and ceased only with death. Although he had an off-hand manner and made friends easily, he did not, according to a former associate, "improve upon acquaintance." Such an easy manner, however, is a necessary ingredient in the make-up of a professional confidence man. Gaining the trust of an individual is the first step in selling a commodity. Harry knew he had the flair and exploited it.

Before long, Orchard was down and out, personally as well as financially. During this time of despair, a friend loaned him enough money to purchase another cheese factory in the town of Hilton. Orchard grabbed at the opportunity, and within the year the friend had been repaid. Surprisingly, business went so well that he established enough bank credit to finance the construction of a new facility near Brighton. To supervise both investments, he managed to stay away from home as much as possible. However, he knew enough to utilize his wife's technical ability to the fullest extent in overseeing the cheese-making process. Since the two plants were located in different communities a few miles apart, Orchard decided to live near the factory that was under construction so that he could personally supervise it.

Relishing his present success, Orchard made a play for his attractive landlady, Hattie Simpson. A few kisses when her spouse was not around, a gentle pat, some words of endearment, culminated in her complete surrender in the master bedroom. Meanwhile, his factory was finished in time for the opening of the cheese-making season. The future looked bright, and in a grand gesture suggesting a touch of conscience, he moved back to live with Florence, who was expecting their child.[3] However, motherhood removed Florence from active participation in the business, making it necessary to hire help, which hurt the company's operation immeasurably.

Much to his wife's dismay, Harry tired quickly of fatherhood and its responsibilities, reacting much as his own father had. The satisfaction he sought was not to be found in the narrow confines of matrimony. His mother's death, which occurred at this time, seemed to push him farther into despondency. Orchard sought solace in liquor and frequent clandestine meetings with his former landlady.

By the fall of 1896, the Lothario of the cheese world found himself deeply in debt again, and credit was nonexistent. The end of his career in the cheese industry was apparent. In a frantic effort to cover up his business activities, he spent long hours of night falsifying the company records. Sensing that something more drastic was neeeded, he searched for an answer that would not only take care of the records but also the debts that plagued him. Since he had already had the foresight to insure his newest company, the final solution seemed to be to burn it.

One lonely night after first removing several hundred dollars worth of cheese products, he put the torch to the building. Reluctantly, the insurance company paid the claims,[4] and overdue bills were paid. At the same time, he made secret arrangements with Hattie to leave Canada and begin the search for western riches.

When Harry Orchard left for Detroit, Michigan, to rendezvous with his lover, he left forever the green rolling hills of his youth. The only possessions he took with him were the pictures of his family. Never returning, he was to die a convicted murderer whose fondest memories were of those he had deserted.[5] But for now, an incredible career lay ahead of him as he set out to find a new life.

CHAPTER 2.

Harry Orchard! That quarrelsome son-of-a-bitch!

Soon after the rendezvous in Detroit, the excitement of the escapade wore off. What were they to do now that the affair had cooled? They considered themselves too deeply involved to turn back. Orchard passed the days playing poker in a third-rate saloon. Hattie, miserable, worried the days away. After two weeks of unhappy living, a vicious quarrel flared, then died amid apologies. The time to move on was near, they told each other. Orchard bought two rail tickets to British Columbia. He was going west at last.

The end of the line for the aberrant pair was the mining district of Nelson, British Columbia. The city was built along the west branch of Kootenay Lake, and because of its mining, lumbering, and transportation importance, it was called the "Queen City of the Kootenays." Orchard was disappointed. Like most mining areas, it was depressing and dismal in appearance, and the snow-covered mountains of the region presented a formidable picture, not at all like the comfortable countryside of Ontario. Nelson was, however, a wide-open enterprise with saloons, dancing halls, gambling dens, and harlots. The town provided the kind of excitement that appealed to a man like Harry Orchard.

The couple set up housekeeping as the newly married Mr. and Mrs. John Little. Cutting another link with his former life, Orchard dropped the name of Albert. Neither bothered to obtain a divorce. Running low on money, they took what remaining funds they had, rented a vacant building, and converted it into a restaurant, a business very much in demand. At first they hired a Chinese cook, but as his culinary ability left much to be desired, he was soon fired. Mrs. Little took over the cooking duties while Mr. Little became waiter, cashier, and dish washer, taking orders with all the finesse of a water buffalo and hating every second of it. Surprisingly, the business flourished.

One night after work while Orchard was hoisting a few drinks in a neighboring saloon, he struck up an acquaintance with a lumberman who was starting a sawmill operation on Pilot Bay, some twenty miles from Nelson. Learning that Orchard was in the food business, the lumberman offered him the job of running a boardinghouse for his lumber crew. He offered to give Orchard the boarding house rent free and to pay Orchard to board his crew. Sucking on the cigar between his thin lips until the tip of it was red with heat, Harry quickly made a decision — to hell with customers. And so the restaurant was sold, and they moved to Pilot Bay.

Arriving at the sawmill site, the Littles hired a full-time cook and Hattie assumed the management of the boardinghouse. Orchard was then free to make a contract to deliver several hundred cords of slab wood to the mill. Hiring two men to help him, he found the hard manual labor gratifying. The operation might have been a big success, had it not been for an unusual circumstance.

Unknown to Orchard, Hattie, in a moment of weakness, loneliness, and uncertainty, had written a lengthy letter to her mother. She responded not only with an impassioned plea for Hattie to return, but also wrote an equally forceful letter to a member of the ministry in Nelson, explaining her wayward daughter's predicament. The minister found the couple at home in the crude boardinghouse and severely chastised them. Did these two immoral people have any feelings, thoughts, or desire for those loved ones they had deserted? Hattie was utterly devastated by his arguments, but Orchard remained unmoved, definitely annoyed, and at the same time unnerved at the ease with which they had been found.

After a sleepless night punctuated with highly charged exchanges about what to do, Hattie, racked with guilt, announced that she was returning home. The next day Orchard rode into Nelson and purchased her a one way train ticket, realizing this was not only the end of an affair, but also the end of his life in Canada. After Hattie left, Harry sold the boardinghouse to the cook, reneged on his wood contract, and paid a fare on the railroad to Spokane, Washington.

Spokane was an ideal place for a man who wanted to start a new kind of life. At that time, it was the foremost trading center for the so-called "Inland Empire," which includes eastern Washington, northern Idaho, Montana west of the Rocky

Mountains, northeastern Oregon, and southern British Columbia. Geographically, the Inland Empire is pretty much of a forested, mountainous area, with multitudinous rivers, streams, and lakes.

In Spokane Orchard found jobs scarce. Ever since the summer of 1893, the United States had been shaken to its foundations by an economic crisis. The depression hit the mining West with a ferocious impact. Particularly damaging to the economic stability of the mining industry was the repeal of the Sherman Silver Purchase Act.

Effects of the depression could be seen and felt everywhere. And the law of the jungle — survival — was pathetically dramatized. Living in a second rate hotel, Orchard stared at the peeling wall paper, the torn, dirty curtains and sun-bleached blinds, and was hit by despondency. His mood matched the melancholy surroundings. In desperation, he wrote a friend in Ontario and received news that his former mistress had returned to her husband. Florence, understandably bitter, was trying to find him, but Orchard had gone too far to turn back, and he knew it. Albert E. Horsley was dead, in more ways than one.

One night, looking for action, Orchard was referred to the well-known Coeur d'Alene Saloon, owned by Dutch Jake Goetz and Harry Baer. The two men had owned a combination saloon and gambling hall in the gold camp at Murray, Idaho, during the 1880s and had given Noah Kellogg, an unemployed carpenter turned prospector, his second grubstake when he was locating the great Bunker Hill and Sullivan mine. For supporting Kellogg, the two received $150,000. Like any good saloon keepers, the pair took their fortune west to Spokane and opened the Coeur d'Alene, a saloon of magnificent proportions.

The three-story building housed several bars, a bowling alley, a theater, a restaurant, and a turkish bath for those who wished to "sober up" and start again. In this miners' rendezvous, Orchard drank and listened to tales of hardships, troubles, and raw camp humor. It seemed to him that half of the miners in the barroom knew each other from one place or another.

Walking down the street one spring morning in 1897, Orchard felt that this was his day. Checking in at the employment office, he came across a help wanted ad for a milk route driver in

Wallace, Idaho. He remembered the years he had hauled milk in Northumberland County. He was experienced and jobs were too scarce for him to be discriminating. So Orchard immediately wrote the Markwell Brothers a letter of application, signing it respectfully, Harry E. Orchard.

Several days later, an affirmative answer to the application was received. After a two-week stay in Spokane, he again boarded a train and steamed across eastern Washington into the scenic beauty and splendor of North Idaho. The train signaled once, and suddenly he was in the fantastic Coeur d'Alene mining district, one of the richest and most extensive in the world. Minutes later, the railway station of Wallace came into view.

National interest in the Coeur d'Alene mining district had been created by the discovery of gold in 1881 by Andrew J. Prichard, who located rich placer deposits near the site of Murray, Idaho, on what came to be known as Prichard's Creek, a tributary of the Coeur d'Alene River. The Northern Pacific Railroad, capitalizing on the find, published a circular letter which greatly exaggerated the mineral wealth and started a stampede to the district in 1883. Ironically, the real lasting bonanza of the region was not to be found in its gold mines, but rather in the rich and extensive veins of silver and lead ores.

The first important silver-lead find was the staking of the Tiger Mine on Canyon Creek at the site of Burke, in May, 1884. Three days later, the Poorman claim was made, situated between Burke and Wallace in Canyon Creek. Practically every silver-lead mine of any importance in the district was discovered in 1884 or in 1885. And the greatest discovery of them all, the Bunker Hill and Sullivan Mine, was made in September, 1885. Noah Kellogg, an unemployed carpenter-turned-prospector, made the big strike, discovering a silver and lead vein of extremely high-grade ore in Milo Gulch. But before Kellogg was through with the Bunker Hill and Sullivan mine, however, he managed to involve it in one of the biggest legal disputes in mining chronicles. Noah had been grubstaked by two different groups of backers, and now that he had made good, each group wanted a share of the find. Final court action gave everyone a piece of the mineral pie.

Now, the development of the Coeur d'Alene district as a deep lode mining producer was assured. Valuable properties matured, and the district gradually attained a more permanent population. As new mines yielded wealth, camps flourished around them. Wallace had no mines in the immediate vicinity, but because of its geographical location, it became the trading and distribution point for large numbers of the camps, while the towns of Wardner, Kellogg, Osburn, Gem, Burke, and Mullan were born to serve adjacent mines.

By 1890, labor problems had arisen. Hard rock mining at best is suitable only for the toughest of individuals. It is hard and demanding labor performed under miserable conditions deep in the "bowels of the earth." The miner is exposed to the possibilities of accidental blasts, fires, or the terrifying cave-in. And the dreaded, undetectable silicosis, the result of fine rock dust inhaled into the lungs, claimed the lives of countless men. Also the miner was often forced to live in company boardinghouses and buy from company stores. On top of all this, constant layoffs and fluctuating wages became a source of bitterness to the miner.

Seeking economic security, miners formed a secret union in 1887, and in August, 1891, they tested their strength and struck the huge Bunker Hill and Sullivan Mine at Wardner, the most prolific mine in the district.[6] Several weeks later on August 29, a satisfactory agreement was reached and the men went back to work. The results of the strike were short lived, for within a week the mine and mill shut down because of a water rights dispute.

Mine owners had viewed the formation of the union with suspicion. Some saw it as a tool of the militant Butte Union of Montana. In any event, the mine owners retaliated by forming their own organization in 1889, the Mine Owners' Protective Association of the Coeur d'Alenes, or more simply, Mine Owners' Association. The strike of 1891 made them more concerned about union activities.

Meanwhile, the mine owners had other problems. Immense capital was required to develop lode mining, and the inaccessability of the region itself retarded development. The substantial amount of money necessary most often came from the financial East, far removed from the action. During the 1890s, the owners were caught between rising operational costs and a fluctuating metal market and finally decided that the only solution left was to cut operating costs by reducing wages.

In 1892, the Northern Pacific and Union Pacific Railroads raised freight rates by two dollars a ton, so to apply pressure, all mines in the Coeur d'Alene were closed from January 16, to April 1. The railroads, losing $5,400 every day the mines were closed, finally agreed to restore the old freight rate. Triumphantly, the mine owners announced that when the mines reopened after April 1, it would be at a reduced wage scale. Enraged, the miners voted to strike, convinced that the railroad incident was being used to break union strength and force them to accept work at any price. The miners refused to return to the mines until they could work at wages which prevailed before the shut-down.

The mine owners had two choices: they could make a settlement with the union or open the mines with imported, non-union labor. They chose to use non-union men. By June, almost a thousand non-union laborers were in the district, and tensions were growing rapidly. Non-union men were accosted and in some cases "persuaded" to leave the area. Businessmen were forced to throw their sympathy on one side or another. The Coeur d'Alene's industries were beginning to come apart at the seams, and a showdown was inevitable.

A violent eruption of armed conflict occurred on the morning of Monday, July 11. The disclosures of Charles A. Siringo's identity on July 9 as a mine owner's spy was the "straw that broke the camel's back." Siringo, a Pinkerton detective, had been hired by the Mine Owners' Association in September of 1891 and had secured work in the Gem Mine under the name of C. Leon Allison. Joining the Gem Union, he was elected recording secretary with access to all union records and activities. Disclosure of his true identity incensed the miners to a fighting pitch, and the saloon brawls that took place on Saturday and Sunday nights were a minor prelude to what happened on Monday.

Early in the morning of July 11, the battle lines were formed, with the large Frisco Mill in the canyon near Gem selected as a target by the miners. After brief exchange of shots, the union people decided to take out the old mill by dynamite. High on the hill behind the mill was a wooden flume that supplied the mine structure with water. Under the direction of George Pettibone, local Justice of the Peace and financial secretary of the Gem Union, water going into the flume was shut off and cases of powder were placed in the penstock which led to the mill's water wheel.[7] The powder boxes slid down the dry slide with tremendous velocity and exploded on impact. The lower portion of the four story mill was razed by the force of the explosion and shattered into wooden splinters. The Frisco people surrendered and were taken to the union hall at Gem.

Fighting was still going on at the Gem Mine, but hearing the explosion at the Frisco Mill and running short of ammunition, the company men gave up, using a formal surrender document. One of the union signers was Peter Breen, a member of the Montana Legislature and Butte Union. At the conclusion of the fighting in the narrow Gem Canyon, five men were dead and many were wounded, including Pettibone.

That night in Wardner, a group of union miners took possession of the huge Bunker Hill and Sullivan concentrator and set up enough powder charges to destroy it. Then an ultimatum was sent to the management: "Get rid of non-union workers or see the mill blown up." The union won its point; non-union men were told to leave the area.

When news of the results of the July 11 battle reached Norman B. Willey, the Governor of Idaho, he asked President Benjamin Harrison for federal troops and on Wednesday, July 13, declared martial law. That same night in Wallace, a riotous celebration was taking place in honor of the occasion of beating the mine owners. The following day a military force entered Wardner, four miles north of Kellogg, and within days fifteen hundred federal troops and Idaho militia men occupied every town in the mining district.

Retaliation was swift and widespread. Arrests of union leaders, members, and sympathizers reached into the hundreds, and since no jail in the district was large enough to accommodate all of the prisoners, stockades or "bull pens" were constructed. These were high fences built around crude frame buildings, with extremely primitive sanitation, food, and sleeping arrangements. The paramount problem which faced the governor was what to do with the prisoners.

Actually, only a handful of people were ever brought to trial.

Of these, thirteen were found guilty of contempt of court and sentenced to several months in the Ada County jail. Four other men, found guilty in a federal court of criminal conspiracy, were sentenced to serve their terms, ranging from fifteen months to two years in the House of Correction at Detroit. By the end of March, 1893, all of the members of the Miners' Union who had been charged with crimes were free. Martial law in the Coeur d'Alenes lasted until November 18, 1892, when tensions relaxed sufficiently.

There is one final bit of irony in the fracas of 1892. When the thirteen men sentenced to the Ada County jail met with their lawyer, James H. Hawley, to discuss labor problems, he suggested that western miners should unite into one organization.[8] It was an appealing idea that started the men thinking and planning. Several months later, at a convention held in Butte, Montana, on May 15, 1893, at the Butte Miners' Union Hall, delegates from the western mining states met and created a new union called the Western Federation of Miners. Just how and when Harry Orchard became involved with this organization is a story in itself.

CHAPTER 3.

A narrow slit in the empty hills of northern Idaho.

In April, 1897, Harry Orchard arrived in the heart of the mining area. Fred Markwell, manager of the dairy, met him as he stepped from the train in Wallace and headed west out of town several miles to his place of business.

"We're pretty much in the center of the district's mining camps and in a good position to serve them all," Markwell remarked.

Orchard looked at the surrounding mountains and fought the feeling of being suffocated by their closeness. He had experienced the same feeling in Nelson, and he wondered aloud if he would ever get used to them.

Markwell laughed. "If this bothers you, wait till you run the Burke milk route." He pointed to a cut in the mountains. "That's Canyon Creek over there. Both Gem and Burke are built on the sides of that narrow canyon. Hell, there's barely room for both a road and a railroad on the canyon floor."

At the dairy, Orchard was introduced to the foreman, August Paulsen, a Danish immigrant who had worked for the dairy since his arrival in Wallace several years before. He had the look of dependability and strong character stamped on his face. As different in makeup as these two men were, they eventually became good friends. It was a friendship that survived for years until Orchard's insatiable greed terminated it. Paulsen showed Orchard around the grounds, explained the daily routine, introduced him to some of the men returning from their routes, and then let him inspect his wagon and gear.

Returning to where Markwell was going over reports, Orchard took a seat and Paulsen left. Markwell put down his papers and looked steadily at Orchard. "Ever been mixed up in any labor disputes?" he asked.

"No," he answered.

"Ever join a union?"

Again the answer was a truthful "No."

At this point, Markwell explained the labor problems that had plagued the district, emphasizing his father had been highly critical of the union activities during the 1892 excitement. The elder Markwell's comments had caused threats to be made against his life in addition to a union boycott against the dairy. The business had been saved only by having the old man leave the Coeur d'Alenes and by selling the business to his sons. Since that time, the dairy had taken a neutral stand on all union affairs.

"My advice to you, Harry, is that regardless of what you see or hear, don't express yourself one way or another," Markwell continued. "If you're ever approached or invited to join the union, do so."

Orchard never forgot Markwell's warning.

Making the milk run up Canyon Creek, Orchard first stopped at Gem, a camp jammed into the canyon bottom and along its walls. Some of the structures still showed evidence of the infamous July 11 gun battle. Then, upon reaching Burke, the end of the line, Orchard was utterly amazed. The rip-roaring camp was built up the sheer walls of the canyon and was even more spectacular than Gem. It seemed as if the buildings were fighting each other to stay aloft. The canyon bottom was so restricted that the one long street almost butted against doorways. The hotel, leaving little room for the ore-hauling trains to pass underneath, was built high across the street.[9] The camp had a reputation for never sleeping, as saloons and gambling places catered to an around-the-clock trade as the mine shifts changed.

Like many mining communities in the area, the school house was incongruously located a short distance away from the red light district. The role that prostitution played was taken for granted and accepted as a natural and integral part of the community. One might say it was an early, western introduction to sex education.

Orchard was fascinated by the vitality of the Burke camp and the clannish attitude of the people, produced in part, no doubt, by its remote and isolated position. He got into the habit, when he finished the canyon run, to have a drink while he observed the play at the gambling tables. Being a stranger did not endear him to the local populace, but with the passing months, he became friendly with bartenders and store clerks.

His exemplary conduct while he first lived in Wallace was soon abandoned. The glass of liquor after work became glasses. He liked rubbing elbows with the rough-talking miners and telling tall tales late into the night. The occasional hand at poker became more frequent. How comfortable it was to slip again into the role he had first played in Saginaw! It was not long before he quit his job and moved to Burke.

In January, 1897, Orchard purchased a small wood and coal business in Burke, rented a cabin, and set out to make a success of the venture. The possibilities were good since the mines, merchants, and homeowners used large quantities of coal and wood. The main requirement for success was consistency in delivering and meeting deadlines. One of his biggest handicaps was lack of operating capital, and like any business it had slack and busy periods. Orchard, by nature, was a poor manager. To carry him during the slack periods, he floated a loan from a clerk working for the Tiger-Poorman Mining Company, but the high rate of interest on the note proved difficult to meet. When the wood and coal business was slow, he occasionally did some packing with horses to isolated mining claims in the hills.

Like most gambling men, Orchard was a natural born loser at games of chance. When he was winning, he would pinch money, carefully and conservatively placing his bets, but when he was losing, he would plunge and hope to recoup his losses in one big hand. He was a loser, and a big one. One night Orchard lost his payroll in a grueling all night poker marathon, ending the night broke and drunk. Not even the charms of a local prostitute could cushion the shock of such a disastrous loss. When he answered a jarring knock at the door the following day, he faced a creditor by the name of Fred Stevens from whom Orchard had been buying his wood.

Stevens got to the point. "Orchard, I got to have the money you owe me."

"I don't have it." Orchard answered lamely.

"You got paid yesterday," countered Stevens.

Seeing no way out, Orchard offered to turn his business over to him, but Stevens refused.

"That won't help my problems. I need money now," he insisted.

Finally, Orchard temporarily pacified the angry man by giving him a mortgage on the business and agreeing to pay him out of the first money he received. The confrontation shook Orchard into a rational course of action. He calmed down, left the cards alone, cut out the big benders, and went to work in earnest. By spring things were looking up, and slowly the business was climbing out of the red.

In the meantime, August Paulsen had presented Orchard with an unusual opportunity. Paulsen was one of a group of local townspeople who had invested money and time in developing a prospect hole into what was hoped would be a paying proposition. The prospect was called the Hercules and was located two miles from Burke on a trail between that community and the old gold camp of Murray to the north. The claim was made in 1889, but the mine had never been a producer and was thought by mining experts to be sterile.[10] Most people thought the only material the hole would give up would be rocks and more rocks, and only the owners had faith in the Hercules.

When Orchard met the owners, the majority of stockholders consisted of Harry Day and his family, Paulsen, the Markwell family, Charles Reeves, and Levi Hutton and his wife May. Reeves was a Wallace barber; the Day family ran a general store under the name of H.L. Day & Sons in Wardner; the Markwell family had the dairy; Paulsen was an employee of the Markwell dairy, and Hutton was an engineer for the Northern Pacific Railroad hauling Burke ore.

These would-be mining magnates had a hard time making ends meet. It was a monthly chore raising money to produce the necessary powder and tools so they could work the mine. For years all the actual digging had to be done on a part-time basis — nights, weekends, holidays — whenever time was available, they were up at the Hercules working.[11]

Orchard liked Paulsen and spent time with him at the mine, questioning his friend as to why he was so certain they would hit that elusive, high grade silver-lead vein. All Paulsen would answer was, "I know it's there; I can feel it."

Sometimes members of the Day family or Charles Reeves worked the mine, but most likely Hutton or Paulsen could be found cutting timbers, setting charges, or mucking out rocks. This was the type of adventure Orchard had heard about in the Michigan lumber camp, and he labored like the rest. When Orchard was offered a one-sixteenth interest in the mine for five hundred dollars, he could not resist. On October 5, 1897, he legally became a partner by purchasing Charles H. Reeves' share.

By this time, Orchard was doing nicely. One payday, as he was making the rounds of saloons seeking men who owed him money, he ran into one of his poker playing friends. After a round of drinks, the man suggested Orchard join him in a few hands of cards at a private club room. The sky was the limit; the ante went as high as twenty dollars, with three raises to a hand. Most pots were counted in the hundreds, and in a few over a thousand dollars were at stake. Orchard lost over four hundred dollars on one hand alone.

Losing every cent he had, but eager to continue, Orchard left the table and looked up Dan Cardoner who operated a grocery store in Burke. Cardoner had been interested in getting a share of the Hercules mine for some time, and at this moment Orchard would have sold his mother's soul to get back into the big game. He offered to sell his interest for one thousand dollars; Cardoner was willing to pay eight hundred in cash. The agreement for the sale was made, and Orchard got his money. But by the end of the evening, his luck had not changed; he was broke.[12]

Wiped out in the poker game, Orchard seemed to lose all interest in business matters, and by summer he was in serious financial trouble. In August, 1898, he had met an old friend from the Michigan lumber camps, James McAlpin, and sold him on the idea that the coal and wood business could be very successful. Orchard was desperate for some ready money to pay off pressing debts. McAlpin had only four hundred dollars cash, but he persuaded his younger brother Archie to work for them, holding his pay until they could get on their feet. McAlpin contributed his savings, assumed half of Orchard's debts, and became part owner of the coal and wood business. Orchard in turn cut his drinking and gambling down to reasonable limits, and when McAlpin moved his family from Michigan, Orchard moved in with them.

Keeping track of his partner, as Jim McAlpin soon found out, had its tedious moments, and more than once Orchard's vicious temper erupted without warning. Orchard was not above lying when the situation suited him, a trait which further strained the relationship. With McAlpin's help though, the stack of overdue bills disappeared, and creditors were even speaking to him again.

Since purchasing wood was their greatest expenditure, the pair reasoned that if they cut enough timber and stockpiled it, they could come out ahead. So in the fall of 1898, while McAlpin stayed in Burke to handle the business end, Orchard moved to Mullan, six miles south of Wallace. Hiring three men, he drove horses to snake out timber out of the thick underbrush. The four men got along well together. When he was not drinking, chasing harlots, or gambling, Orchard could be a considerate and thoughtful employer.

For entertainment, Harry spent his spare time in Flynn's Cigar Store in Mullan. Dominick Flynn, the owner, had a club in connection with the cigar store. Orchard would come in, draw a brew from Pat McHale, the bartender, and at the first opportunity would join a nearby card game.[13] Orchard felt comfortable at Flynn's.

In February, 1899, Orchard shut down wood cutting operations and moved back to Burke. Winter weather was hampering the work, and the expenses were fast becoming greater than the timber's value. The following month Orchard got into another big poker game and lost everything he had. Exactly twelve months before he had lost his mine interest and this time his business. Orchard was a born loser.

He woke up McAlpin, and the two men talked the night away. After breakfast Orchard turned the entire business over to his partner.

"Are you broke?" asked Jim.

"I don't have enough money in my pockets to buy a meal," Orchard replied.

McAlpin gave his friend a hundred dollars and agreed to settle his bills. Two years earlier Orchard had bought into the business for one hundred and fifty dollars. One month later, he went to work at the huge Tiger-Poorman mine at Burke. Lewis Strow, a shift boss and drinking partner, helped him obtain the job of "mucker," a person who shovels out rock after it has been blasted. Orchard had joined the "Knights of Labor" when he first purchased the wood and coal yard; now he had to effect a transfer to the Western Federation of Miners. And it was this new affiliation that was to change his life so drastically.

CHAPTER 4.

Every union should have a rifle club.
. . . Edward Boyce

It was hard to avoid the union, so powerful was its control of the region.

After the organization of the Western Federation of Miners in 1893, it had been easy to bring many existing unions into the Federation. Prior to its founding, there had been in the various mining camps of the Rocky Mountains, unions or organizations which represented the miner's interests. These locals, for the most part, usually acted independently of each other except in certain sections of the country where several unions formed a central union composed of delegates from the local unions.[14]

The internal organization of the Western Federation of Miners consisted of an executive board of five members and the president and secretary of the union. This board was given almost absolute governing power except during the few days when the annual convention of delegates was in session. They were given the authority to issue orders, carry out policies, levy and collect assessments, and use federation funds. The so-called "Inner Circle" of the federation was composed of a few select men from the executive board who actually ran the organization, and often the entire district.[15]

A year after the founding of the federation, that body began to forcefully test its strength. Violent and bloody strikes occurred in many different mining districts in the western states: Cripple Creek, Colorado, 1894; Leadville, Colorado, 1896; Coeur d'Alene, Idaho, 1899; Telluride, Colorado, 1901; and Cripple Creek, Colorado, 1903-04. These strikes extended to other camps, and from 1901 to 1904 there was almost constant turmoil in the mining camps of Colorado.

Edward Boyce, president of the Western Federation of Miners, officially set the policy for his union. Boyce worked in the famed Anaconda mine at Butte, Montana, then drifted into the Coeur d'Alenes, where he mined and became a leader in the infant union. As a result of the 1892 troubles, Boyce was one of those sentenced to the Ada County jail in Boise, from July, 1892, until

February, 1893. When the Western Federation of Miners was born on May 15, 1893, he was one of the men so instrumental in getting the union started.

Boyce was president of the Federation from May, 1896 to May, 1902. Under his leadership and direction, union membership increased from seven or eight thousand to about thirty thousand. On May 10, 1897, Boyce addressed the annual convention of the Western Federation of Miners, assembled at Salt Lake City, Utah. His speech answered many questions concerning the direction the federation was going:

> Every union should have a rifle club. I strongly advise you to provide every member with the latest improved rifle which can be obtained from the factory at a nominal price. I entreat you to take action on this important question, so that in two years we can hear the inspiring music of the martial tread of 25,000 armed men in the ranks of labor. I would recommend the adoption of a new ritual. The constitution (of the Western Federation of Miners) should also be amended to declare all members of the National Guard ineligible to membership and to withhold our patronage and assistance where any member of the National Guard is employed or admitted to membership.

Five days after this militant speech was delivered, the building housing the guns and the ammunition for the state militia at Mullan, Idaho, was broken into and weapons were stolen. The federation always denied being involved in the incident, but by a strange coincidence, firearms turned up in the hands of union members in 1899. The Boyce rifle club of the Coeur d'Alenes had entered the contest.

Blood was spilled again on December 23, 1897. This time, a party of masked men shot Frederick Whitney, a non-union foreman of the Frisco Mill. Whitney was taken from his rooming house in Gem late that night and was shot when he tried to escape. Taken to the hospital at Wallace, he died on Christmas Day. The slayers were never apprehended.

Control of the Coeur d'Alenes was further tightened by shrewdly using the power of the ballot box. Public officials sympathetic to union causes were elected at every opportunity.

Particularly important to their efforts were the law officers and local justices of the peace. Whenever a non-union laborer got into trouble, he was taken before the justice of the peace and was fined, jailed, or advised to leave this district. Orchard said that Pettibone, justice of the peace at Gem in 1892, used to boast how he had used his office to ship in rifles to the union from Denver in 1899. He claimed to have sent in hundreds of them in piano boxes, addressed directly to the sheriff.[16]

By the time the labor dispute of 1899 blew the Coeur d'Alenes apart with its savagery, the mining district had become a human powder keg of emotions. After the riots of 1892, a temporary truce existed between the federation and the various mining concerns. However, it had developed into a very sensitive situation on the verge of exploding.

The final eruption occurred due to the persistent efforts of the federation to unionize the Bunker Hill and Sullivan mine. The company had raised its wages to union scale, but it still refused to recognize the federation's power or meet the demands that non-union employees be discharged. Since the Bunker Hill and Sullivan was the largest, wealthiest, and most influential property in the district, and because of its resistance, it was felt it must meet the federation demands or be destroyed. Nonetheless, the mining company, under the direction of its general manager, Frederick Bradley, stood fast. Enraged, the Federation decided the company must be taught a lesson, and Bradley especially must be made to pay.

Orchard was on his way to a small restaurant on that April day, when an acquaintance stopped him and announced that a special meeting had been called for 7:00 a.m. at the Burke Miner's Union Hall. All members were required to attend. The importance of the meeting could not be misunderstood since there would be no work at any of the mines that day. Harry rushed over to the Union Hall, sensing that something big was going to happen. The gathering crowd quickened the feeling of excitement. Filing into the hall, he chose a seat as close to the speaker's rostrum as he could.

The meeting was called to order by the secretary of the Burke Miner's Union, Paul Corcoran, who was also a representative to the Central Union. A caucus of the Central Union had been held the night before at Gem, and the decision had been for federation members to go down to Wardner in a body and destroy the huge Bunker Hill and Sullivan mill. Corcoran's voice raised emotionally as he stressed that other mines in the district, with the exception of the Bunker Hill, had temporarily granted the demands of the Federation and would go along with any agreement made with the Bunker Hill and Sullivan.

What really fanned the Federation's fire was the determined attitude of the Bunker Hill management not to acknowledge the union. Union men were leaving the Federation, and non-union laborers that had been driven out of the district were returning. The solution as offered by the Central Union was to go down to Wardner en masse, drive the scabs from the Bunker Hill, and then destroy the mill.

According to Harry Orchard, the following plan, presented by Paul Corcoran, had been agreed to by the Central Union:

> The Northern Pacific train left Wallace for Burke every morning about ten o'clock, and he (Corcoran) said after the train left Wallace, the wires would be cut so that the railroad officials could not find out anything that was going on in the canyon. He said when the train arrived at Burke, they wanted to be all ready, and five or six men would go out and take possession of the train and give the train crew orders to make up what empty cars were needed and turn toward Wardner. He said the Gem union would be ready when they arrived at Gem, and the Mullan unions would meet them at Wallace. He said it had been arranged to stop at Frisco Magazine, and it would be open, and the would load on what powder they wanted to blow up the mill, and he wanted everybody that had any firearms and ammunition to get them. He said they had some new rifles, and they would be distributed at Gem, and he thought they would have enough firearms to easily whip the company of militia.

Corcoran closed the discussion by stressing that the plan was the desire of the Central Union and that the governor would not take action since the district had supported him for office. A final

vote on the proposal was taken, and the plan carried by a small majority. With that, nearly every miner in Burke prepared for the trip to Wardner. The men were told to tie a piece of white material around an arm to signify that they belonged to the Burke union, as well as to devise some kind of mask for disguise. The wearing of the masks was a macabre touch, reminiscent of other times in the district when secret groups had met to kill.

Meanwhile, the action that was taking place in Burke was also being enacted in other mining camps of the district. At Mullan, union members hurried to an old barn and dug up boxes of rifles which had been buried under refuse. At Gem, masks had been made and guns had been arranged for. And so the mass movement was proceeding with utmost military precision.

After the meeting, Orchard had returned to his room, opened his luggage case, removed a revolver, and shoved it into his waistband. He stopped for a moment to wipe perspiring hands onto his pants, then hurried down the stairs and continued out of the building and along the narrow wooden sidewalk to the waiting train, excited by the prospect of the impending conflict. When he arrived at the station, the Northern Pacific was under the command of armed, masked men. Telegraph wires had been cut and the railroad crew had been ordered to put on extra cars.

The Northern Pacific first stopped at the Frisco Magazine, located a little above the town of Gem. Here, forty fifty-pound boxes of powder were loaded. Then the train continued slowly into Gem where it stopped again at the Miners' Union Hall. Now, more weapons and ammunition were distributed among the men already on the train. Union leaders apparently had second thoughts about the amount of explosives needed to level the big mill, and they ordered the train to back up the canyon to the Frisco Magazine, where another load of powder was added. After taking on the miners from Gem, the motley army of dissatisfied men rolled west to Wallace. Sounds of guns being loaded, nervous coughs, spasmodic laughter, and excited conversation filled the air as they moved toward their destination.

Arriving at Wallace, union men from Mullan and Murray joined the entourage. Also boarding the train were Sheriff Jim Young and his deputy, Tom Honey. Orchard remarked, "I was told they were advising the men on the way down to Wardner how to do the work and not get into trouble over it." In any event, by this time hundreds of men were on the train, crowded inside the cars, hanging on the sides, sitting on the top, hungering for battle.

At the Wardner depot, the Northern Pacific was greeted by a large crowd of federation men. As the men disembarked from the train, the sheriff stood in front of the group and demanded that they disperse and return to their homes. Everyone knew this was a pretense, but it was something that was legally demanded of him. Laughingly, the miners walked past him and began forming for the attack. William F. Davis, the Gem Union leader, seemed to be in charge. The miners were lined up according to the type of weapon and the union they belonged to. An estimated thousand men, two by two, marched toward the mill half a mile away. The marching feet of the union rifle club that Boyce had wanted were on their way at last.

As the miners moved out to attack, Orchard remained at the depot restaurant to try to get something to eat. "I had only a small revolver anyway and wouldn't have been any particular use," he later claimed. The truth seemed to be that he had more stomach for food than for the fight. In a few minutes the sound of gunfire was heard, and the depot emptied as the occupants ran out to see the fireworks.

After showing a token resistance, the mill was taken by the federation. The powder was unloaded from the railroad car, and a large group of miners hanging around the depot were pressed into service as bearers. Orchard carried a fifty-pound box on his shoulder. Arriving at the mill, the powder carriers found Davis in charge of the explosive placement, which he had planned with all the skill of a professional.

Explosives in place, volunteers were called to fire them off. Orchard had been fascinated with the preparations. And the prospect of seeing the big mill go sky high was so alluring he was more than willing to go back into the building and light the fuses. Orchard and another miner entered the mill, went down into the boiler room, and stood next to the stacked dynamite. Then Davis thrust his head through a trap door and yelled for the men to light their fuses. The fuses burned with a sputtering speed, and sparks flew in a circular motion, leaving a heavy black smoke in their wake. A moment of panic came when the pair prepared to leave the doomed structure only to find the door of the stairway locked. Back down the stairway they stumbled and luckily found a window. Pulling themselves through, they ran from the building and across a switch track to safety behind freight cars standing in the yard.

Crouching behind the huge iron cars, they watched as the charges exploded fifteen seconds apart and blew the quarter million dollar mill to pieces. The huge and imposing structure was reduced to a mass of kindling wood and twisted metal. Prior to the spectacular blast, other buildings, including the superintendent's house, the office, and the company boarding house, had been set on fire. Now, the mill was a pile of rubbish, and the smell of kerosene permeated the air as other buildings burned. The crowd cheered and shouted; guns were discharged in salute. The waiting train was reloaded with the victorious warriors, and it started back toward Wallace and Burke.

Halting at Wallace, federation men from Mullan and the immediate area got off the train. Bottles of liquor were opened and handed around. Orchard took a drink and passed it to the next man. Not even rumors of federal troops coming into the district subdued their spirits. The train continued, picking up speed. Orchard held on with both hands as the wind cut his face. He later wrote, "That evening I went back home and went to bed as usual without thinking much about it." He was not the emotional type.[17]

The next day, most of the miners reported for work as if nothing unusual had taken place. Each day the stories that troops were on the way became more prevalent among the miners, and as the days passed, fewer and fewer men were showing up for work. At this point, the full impact of what would happen if troops did appear finally penetrated Orchard's thinking. Just before ten o'clock, he quit the Tiger-Poorman and picked up his time, but he was unable to get paid that day.[18] He was fully aware now that time was of the utmost essence. Checking out of his room, he took his luggage to a friend's house and left it.

Orchard started hiking over the mountain on the north side of Burke. There he found others in the same situation — Jack

Collins, Carl Emerson, Harry Gordon, Patty Dennison, and many more. By late afternoon, the troop train was churning up the canyon. When night came, two men from the group crept silently into town to find out what was going on. They found the camp alive with soldiers, intent on arresting every man, regardless of profession or business.[19]

Late into the night, the men discussed their dilemma and the choices left open to them. Most of them decided to make the journey over the rugged Bitteroot mountains to Thompson Falls, Montana, about forty miles away. Early the next morning a count of those wanting to cross over to Thompson Falls revealed only two, Harry Orchard and Pat Dennison, and at five o'clock a.m. they left.

The snow was deep, but they made good time on the hard crust. As the sun thawed the snow, the men sank in up to their waists in some areas, but still kept moving. Ten o'clock that night, they trudged wearily into Thompson Falls, hoping to elude the federal troops.

CHAPTER 5.

I say to you that the organization which hides from the open light of the day, whose members flee before the dawn, "like a guilty thing upon a fearful summons," is not an organization which can represent the dignity, the manhood and patriotism of American labor or the Western miner . . .
. . . William E. Borah

Harry Orchard left not a moment too soon. Federal troops under the command of Brigadier General H.C. Merriam arrived in the Coeur d'Alenes on May 2. A personal representative of the governor, Bartlett Sinclair, the state auditor, had arrived in the district the day before to investigate the situation first hand. He declared the Coeur d'Alenes to be in a state of insurrection and rebellion and imposed martial law on May 3.

Men were arrested, but many like Orchard had already escaped. Sinclair, liberally using powers not found in Idaho state law, began his round-up. This action was too reminiscent of 1892 not to send a shudder of fear throughout the mining district.

If any one thing could arouse the indignation of the civilian populace, it was the infamous "bull-pen," which served as an internment camp. At first the "bull-pen" consisted of an old barn and a couple of box cars, with hay spread on the floor for sleeping purposes and the major problem, sanitation. Later in the month, the prisoners finished building a new military prison, and the men were transferred there.

In a final effort to rid the Coeur d'Alenes of Federation men, Sinclair set up the permit system for mine workers. In order to secure a permit, the miner had to sign a statement declaring that he had not taken part in the activities of April 29, and that he renounced his membership in the Western Federation of Miners.[20] As expected, the permit system aroused immediate protests from labor groups throughout the country. Sinclair pushed the permit system over the objections of the mineowners, and the blacklist worked.

The crux of the state legal action centered around the killing of the non-union miner, James Cheyne. Among those indicted for the murder and conspiracy was the secretary of the Burke Union,

Paul Corcoran, who was placed on trial July 8. George H. Stewart, a Boise district court judge, was selected as trial judge. The prosecution consisted of Jame H. Hawley, who was formerly defense attorney for the miners in the 1892 trials. William E. Borah also of Boise and J. H. Forney of Moscow were also retained by the state. The defense was ably represented by Colonel Pat Reddy of San Francisco, a former partner of Hawley in the 1892 trials, F.C. Robertson of Spokane, Peter Breen of Butte, an active participant in the 1892 riots, and Walter Jones of Wallace.

The closing argument for the state was delivered by Borah, when he said:

> We ask you, therefore, was there a conspiracy? Not that these men got together and deliberately entered into a partnership, as counsel for the defense say they must have done, or signed an agreement to do a certain thing. The court will instruct you that it is not necessary that there be any agreement between them; it is only necessary that they act with a common and ultimate object in view. In this case, we find these men and the unions, although many miles apart, almost at the same hour of the day, acting apparently with the same purpose in view, with the same design, making the same preparations, moving in the same direction, and arriving finally at the same destination. How did it happen? No, no, you must be satisfied that it was all understood. That between the members of these unions and especially the officers there was a positive understanding that upon the 29th day of April every union should gather its men together, provide them with arms and masks and send them forth for the purpose of destroying the Bunker Hill and Sullivan mill and to drive out the non-union employees. They all arrived at the same point by the same train, at the same time, and when they got there, from that moment, they became, in the eye of the law, one individual; the act of one was the act of all. I ask you to bear in mind the startling fact that there was scarcely a hitch in this movement from the time it began until they returned after the crime had been committed upon the night of the 29th.

To prove that the defendant was a member of the conspiracy, Borah cited Corcoran's position as secretary of the Burke Union, as well as his being a delegate to the Central Union. To further implicate Corcoran, Borah related an incident that happened on the morning of April 29, when Mr. Culbertson, manager of the Tiger-Poorman mine discovered that none of his men had reported for work. Culbertson sought out Corcoran, who replied, "There is going to be no trouble in the canyon. We are going to Wardner today."

Borah stressed that, "The force he set in motion there at Burke was the force which took the life of James Cheyne," and gave this final plea to the jury:

> It is for you to say what our young state shall do. Shall high handed crime continue within her midst and ply its trade in open defiance of law? Is our young state to become the rendezvous for criminals, the by-word of sister states? Is our state's pride forfeited? Is our manhood dead? I appeal to you, as men and citizens, to give back the reign of law; deal fairly but fearlessly with those who would continually trample all authority and the state's honor beneath the feet of lawless vengeance.
>
> To you I now submit the whole cause, and may the power which works for the betterment of all give unto you the righteousness of judgment which will enable you to deal in justice and without fear of man or the dread of man between this defendant and your sovereign State.
>
> I thank you again and again for your attention and the exercise of your patience and submit this matter for its final adjudication in the court of your own conscience.

At the conclusion of Borah's address, Judge Stewart delivered his charge to the jury, reminding them of the defendant's alibi in relation to the conspiracy question and offering a clarification statement.[21]

> One of the defenses interposed in this case is what is known in law as an alibi, and it is a proper and legitimate defense when proven, and consists in showing that at the

> time the alleged crime was committed the defendant was at a place different from that where the crime occurred. And where the defendant relies upon an alibi for his defense, the burden of establishing such alibi is upon the defendant, and if he succeeds in raising a reasonable doubt in the minds of the jury as to the fact of his presence at the place and at the time the offense was committed, he is entitled to an acquittal, as it is incumbent upon the prosecution to establish the guilt beyond a reasonable doubt. An alibi, however, is no defense in this case, if you should find a conspiracy existed and that the defendant was a member thereof, as stated in instructions.

Corcoran was subsequently found guilty, and Judge Stewart pronounced the sentence of seventeen years of hard labor in the state penitentiary. The murder trial was over.

Very few of the men confined to the "bull-pen" were ever brought to trial. Eight other union leaders were also going to be brought to trial on the conspiracy charge, but they managed a successful escape and never faced charges. Ten men, however, were sentenced in federal court to serve sentences of two years for interferring with United States mail that was on the Northern Pacific train making the forced run to Wardner.

The labor troubles of the Coeur d'Alenes finished Governor Frank Steunenberg politically, much in the same manner it had ruined former Governor Norman Willey after the 1892 riots. Bitter feelings toward Governor Steunenberg had existed for years, and he had received anonymous, threatening letters. Five years after he left office, he was assassinated.

Meanwhile, in another part of the state, at Silver City, a dirt-crusted miner who was working deep underground in the Florida tunnel of the Blaine mine, shut off the air on his heavy power drill. In the dim light, he picked up an empty powder box and on its side wrote a resolution condemning Governor Steunenberg for his actions in the Coeur d'Alenes. Later that night he transferred it to paper, and at a meeting of the Silver City Union, number 66, of the Western Federation of Miners, he introduced the resolution to its members. It was adopted and subsequently printed in the union's publication, *Miners'*

Magazine.

Still worked up over the injustices he believed were apparent in the handling of the Coeur d'Alene labor problems, the man remarked to W.T. Stuart, an engineer for the Blaine mine and mill, "Steunenberg ought to be exterminated." Stuart was taken aback by the statement coming from Bill Haywood whom he considered one of the best citizens in the camp.

At the same time, Frederick W. Bradley, manager of the Bunker Hill and Sullivan mine was both weary of and wary of the mining district. Being directly involved in two large scale labor wars was more than enough. In the minds of Federation people, he was as much responsible for their misfortunes as was the governor. He moved his operations far away to the city of San Francisco, but he was not forgotten by Federation leaders.

Peter Breen, dynamiter and lawyer, returned to Butte where he was elected prosecuting attorney. Jack Simpkins had received some harsh treatment in the "bull-pen" during his confinement, but he remained committed to the Federation. Released, he continued his union connections and rose to a position on the Federation Executive Committee, becoming a member of the influential "Inner Circle."

The power and influence of the Federation had come to a halt in North Idaho. The union gave up the Coeur d'Alenes, but decided that the mining districts of Colorado were ripe for a concentration of Federation efforts. Moving its headquarters from Butte to Denver, the union was located in the Mining Exchange Building of that city. Idaho was quiet at last, but what of Colorado?

As the court battles raged, Orchard and Dennis moved on. In Thompson Falls, they found other displaced miners. Rumors were circulating that federal troops were on the way to pick up stragglers who had escaped over the mountains. Realizing there was probably a measure of truth in the stories, the weary miners drank cups of strong, black coffee, caught a few hours of needed sleep, and were on the three o'clock train for Missoula, Montana.

At Missoula the pair faced the same problems. Troops were supposedly on the way and had deputized local Coeur d'Alene men who could identify the escaping Federation miners. For

Orchard and a dozen other fugitives, the solution to their problem seemed to lie in the Bitterroot Valley, south of Missoula. There the group stayed at the farm of a friend, but a few days later Orchard and another man made their way to the farm of Pat McCarthy, who had formerly driven a milk wagon for Markwell Brothers.

Explaining his predicament with all the persuasion of a man about to go to prison, Orchard finally convinced McCarthy to make a trip into the battle-ripped Coeur d'Alenes to collect the money the two men had due and also to pick up their stored baggage. In return, the men agreed to work the farm in his absence.

McCarthy encountered no difficulties in obtaining either the money or the baggage, and he also gave the men a rundown on what was happening in the mining district. Two things particularly impressed him — the presence of the federal troops and the notorious "bull-pen." The thought of being confronted by either of these was enough to make Orchard further appreciate his escape into Montana. He spent another ten days in the Bitterroot Valley working for a rancher named Briggs for board and room. Being a ranch hand, however, was not Harry's idea of living, and he knew it was time to move on.

This interlude in the Bitterroot Valley points out again Orchard's strongest characteristic — his ability and his desire to exploit people. In this instance, he had talked a mere acquaintance into making a trip to Burke, under adverse conditions to retrieve his luggage and his pay checks. Orchard had felt no qualms in asking McCarthy to travel several hundred miles for his own personal benefit.

When Harry Orchard left the security of this rural community, he picked up with his restless friends and returned to Missoula to assess their predicament. No soldiers were in town, so they felt comparatively safe. The consensus of the group was that they should go into Butte, headquarters of the Western Federation of Miners and the location of some of the world's greatest copper mines. Orchard became ill on the ride to Butte and did not completely recover for several weeks.

Butte was booming as never before, and in the process the town acquired various reputations: a drinking man's town, the best whorehouses west of St. Louis, the strongest union town in the West, and an abnormal death rate in the deep mines.

Orchard and his friends proceeded directly to Federation headquarters and obtained a withdrawal card from that union so that they could gain admittance into another mining organization if necessary. Ed Boyce talked to the group and urged them to return to the Coeur d'Alenes as soon as the troops were removed.

Orchard stayed in Butte about a month, waiting for the trouble in Idaho to be settled. He learned that the mines were operating again, that the permit system had been adopted, and that the district was still under the jurisdiction of the military. Representatives of the Standard and Mammoth mines came to Butte and attempted to recruit five hundred men, offering union wages and train fare to the district. Since they asked the Butte union to secure the needed manpower, a meeting was held at the Union Hall to consider the matter.

After hearing the proposition of the representatives of the Standard and Mammoth mines, the union members decided that more trouble was inevitable because of the stringent permit system, which was totally unacceptable to them. The butte union turned down the offer, so the company representatives left for Joplin, Missouri, a non-union camp, where they were certain to be more successful.

It was now evident that an immediate return to North Idaho was out of the question. Nearly broke, sick, and observing the large number of unemployed men in the same position, Orchard began to do some serious thinking and finally realized there was no hope here. Down to his last seventy-five cents, he entered one of the many wide open gambling establishments, and with a rare streak of luck ran it up to three hundred dollars. He had his stake.

Orchard bought a one-way rail ticket to Salt Lake City, Utah. Utah, he had been told, was actively hiring miners, and it was far removed from troubled North Idaho. Still recovering from his illness, Orchard rested in Salt Lake City for two days and then headed for the mining town of Bingham, a short distance away. This town, he found to his delight, was similar in many ways to his former stamping ground of Burke. Bingham, too, was built along one street at the bottom of a narrow canyon and was also a wild and rowdy place. Thirty saloons catered to a drinking man's thirst twenty-four hours a day. Gambling was a common activity, and the predominantly male population consistently supported the many houses of prostitution.

It was not until the first part of the twentieth century that Bingham's copper ore, low grade sulphide containing less than two and one half percent, was successfully mined. Although men were aware of the existence of copper ore, it wasn't a profitable venture until Daniel C. Jackling, an engineering genius, whose theories of open pit mining and the treatment of low grade copper ore made mining the ore both possible and highly profitable.

The wandering union fugitive found work in the Ben Butler mine and later in a prospect, working for a contractor who was intent on driving a tunnel into worthless earth. One of the surprises he encountered in the Bingham camp was the association with the numerous men he had known in the Coeur d'Alenes. Most of them were using assumed names since word had spread that the Mine Owners' Association had circulated a black-list of miners who had worked in the Idaho district up to the time of April 29. Orchard himself had been using the name of Dempsey. Since there was no union in Bingham at that time, he did not have to decide whether or not to associate himself with it, so he went directly to work with little or no formality.

Orchard spent the Fourth of July, 1899, in Salt Lake City, drinking and gambling. The following morning he was in the saloon nursing a miserable hangover when he heard of some contractors who were sinking a shaft in Little Cottonwood Canyon, just out of the Salt Lake City limits. He was employed until Christmas.[22]

During the winter months of 1900, Orchard was in the St. James mine in Butterfield Canyon, near Bingham, making two dollars a day in wages and another two dollars in mining stock. Although he was constantly short of cash, he always indulged in personal pleasures. In the spring he returned to Salt Lake City and found enough employment to supply money for drinking.

In November, 1900, he took off for the fabled San Francisco, with no plan in mind. To keep on the move was the important thing. Like others, Orchard found San Francisco a good town for night life, and he spent money on his usual amusements: liquor, poker, and women. He was reluctant to leave the big city. His next trip to San Francisco would be to kill; but for now, his stay was for sheer pleasure.

Leaving the Gate City, he traveled north to Lake County where he obtained a contract to provide fire wood for the Bartlett Springs Company. This job provided a steady income. However, spring stirred his restive nature, and he left for southern California in May, 1901. During the two weeks he was in Los Angeles, he spent much of his time with an old mining acquaintance by the name of Grant whom he had accidentally run into. It was pleasant to meet Grant again, but as usual, nothing could really hold him for any length of time. He moved on.

Arriving in Salt Lake City, Orchard found a job driving a milk wagon for the Keystone Dairy. The delivery run was a far cry from his hectic route to Burke, and by fall the job was downright boring. There must be something better to do! Orchard soon found a change of pace in Arizona, driving a freight wagon for a mining concern. This adventure lasted through the winter months until March, 1902. He was now making three dollars and fifty cents a day, and despite his usual excesses, he managed to save two hundred and forty dollars. With that much money in savings, he decided it was time to take leave of Arizona.

Orchard returned to Salt Lake City and stayed just long enough to exchange greetings and the latest news with old cronies. He then journeyed out to Bingham Canyon to see more friends and to promote a poker game. Once again he experienced a losing streak, but this time he quit the game while he still had a hundred dollars in his pocket.

When he returned to work, it was at the Ophir mine at State Line, Nevada. There he worked twenty shifts in the quartz mine and earned four dollars daily. When he left Nevada, Orchard had only fifty dollars; the rest of his earnings had been left on card tables and bar counters. This time, he returned to Salt Lake City.

On his trip to Nevada, Orchard had been accompanied by Arthur Doolan, a man he had met in a Salt Lake City gambling place in 1899.[23] At Doolan's request, Orchard took a short prospecting trip into southern Idaho, near Paris, but that lasted only a few days. He talked fluently after a few beers, but in the mining field his ability was limited to the manual tasks of a hard rock miner. He was not a consultant. In any event, shortly after July 4, 1902, Orchard and Doolan left for Cripple Creek, Colorado, "The Greatest Gold Camp in the World."

From the time of the Bunker Hill and Sullivan mill explosion until the time Orchard departed for the Colorado gold fields, he was a wanderer, a hobo, a transient, a man without purpose or direction. He worked most of the time and made good money, but he spent it before it got cold in his pocket. On the move most of the time, he spent much of his travel time "riding the rails." Orchard usually had the money, but he enjoyed the excitement of a free ride without being apprehended. Occasionally, he did get caught, but always he managed to bribe his way out of difficulty. He found an added thrill in this, also.

Writing about this period, Harry Orchard stated that he had considered changing his manner of living many times:

> I made many good resolutions, and often saved up a few hundred dollars and thought I would get into some little business for myself. When I could get away from town, as I often did, in some out-of-the-way place, I would save my money and make good resolutions; but how soon I would forget them when I would strike town and see a faro game running, or a game of poker; my money would burn my pocket. There were many other attractions and money always soon got away.

There were too many distractions in Orchard's style of life for a man who had such a low resistance to temptation. Later, he remarked, "I always bought plenty of good clothes and lived well." At any rate, Cripple Creek was to be the turning point for Harry Orchard.

CHAPTER 6.

I thought it would kill a cage-load of non-union men, as the men always went down first going on shift.
. . . Harry Orchard

Cripple Creek was where the action was in 1902. Miners, gamblers, whores, speculators — all were after whatever gold could bring them, and more. Orchard and Doolan arrived at Cripple Creek about the middle of July. They first stopped at Victor, but after a few days they moved to Independence. Orchard was unduly impressed with the rich mining district that boasted almost five hundred mines and several booming towns. He wrote:

> It was more like living in a city than a mining-camp. They have a fine opera-house at Victor and also one at Cripple Creek, and nearly all the good plays come there. There are good hotels. There are no company boarding-houses or stores. All work at the mines is eight hours. The wages run from $3 to $4 per day, and without exception this is the finest mining camp to work at in the country, if not the world.

For the first two or three weeks in the mining district, the pair batched together. Orchard got a job in the Trachyte mine operating a machine drill, but Doolan could not or would not find work. As the days passed, their friendship fell apart; the fact that Orchard had to bear the expenses for both of them was too much for him, and soon they parted company. Doolan left for Victor to find employment, and they saw little of each other after that.

Before the split occurred, however, Doolan introduced Orchard to John Neville, a saloon owner. Neville, a Canadian by birth, was about forty-nine. He had been a miner, but because of an injury in a mine accident, he had been forced to find other work. And what better business is there in a mining camp than owning a saloon? The two became good friends. Orchard found him liberal and kind-hearted, and after Doolan left, Orchard boarded awhile with the Neville family.

Orchard worked as a miner and earned four dollars a day. His room and board cost him only thirty dollars a month, but he was always broke at the end of the month. Most of his money was spent for liquor at Neville's saloon. There were no regular card games, but usually a game called freeze-out was played for drinks. An opportunity to play cards and drink was always difficult to resist; consequently, Neville eventually got most of Orchard's pay check. It was Neville who introduced Orchard to some of the mine managers, resulting in a job at the Trachyte mine.

Orchard's job at the Trachyte was as a contract worker with another man. When the contract ran out in November, he managed to land another job at the Hull City Placer where he timbered and operated a machine drill. In March, Orchard's contractor, Mr. Warren, moved over to the Vindicator No. 1 mine to handle some piece work mining, and Orchard went with him, remaining there until the strike of August, 1903. The Vindicator No. 1 mine was situated on the infamous Bull Hill, a little way down from the Hull City mine, site of a shattering strike in 1894. The seventh, eighth, ninth, tenth, eleventh, and twelfth levels were being worked in the Vindicator. Orchard operated a drill on the seventh and twelfth levels.

During this period, three important events occurred in the life of Harry Orchard. The first was his realignment with the Western Federation of Miners. After he left the Coeur d'Alenes, he had never worked in any union mining camp until he arrived at Cripple Creek. A week after he began working at the Trachyte mine, William F. Davis and William B. Easterly asked him to join the Free Coinage Miners Union No. 19 of Altman.[24] Orchard recognized Davis as the man who was in command of the gang that blew up the Bunker Hill and Sullivan mill. Davis had since moved to Cripple Creek and was president of the Altman Union, while Easterly was secretary.[25] Orchard explained to them that he had been a union member at Burke, but had lost the withdrawal card that Boyce had issued him in Butte. After the first payday, Orchard promptly rejoined the Federation and became a member of the infamous Altman Union.

There were eight unions in the Cripple Creek mining district:

one each in Cripple Creek, Anaconda, Altman, and Victor; an Engineers' Union at Cripple Creek and at Victor, and a Mill and Smelters' Union at Victor. Each local union selected one or two delegates who composed the District Union.

The largest union was at Victor, but by far the most radical was the Free Coinage Union at Altman, which was labeled "the Bull Hill dynamiters," and had the immediate area of Bull Hill under its control. In the larger towns such as Cripple Creek, some of the mines still employed miners who did not belong to the union. Strong union pressure in the Bull Hill area accounts for Orchard's joining the Federation the second time. Until trouble arose, however, he showed little interest in union affairs, and then it seemed he had joined only for the excitement, the money, and the feeling of importance.[26]

The next event that affected Orchard was the act of getting married, even though it was illegal and consequently made him a bigamist. Despite his addiction to liquor and cards, he often suffered the pangs of loneliness especially in the quiet, late hours of the night when he thought occasionally of his wife and child whom he had deserted. "I would think of my dear wife and little girl, and wonder if they were still living and how they were getting along." Despite such feelings, he never attempted to correspond or to inquire about them, and with the aid of more liquor he managed to push such thoughts from his mind.

Most of the time the girls of the red light district took care of the ruddy faced miner's physical need, yet the security of a home was still a dream. Perhaps contentment would come if he had a real home to return to at the end of the day, he began to tell himself. In the spring of 1903, he met a pleasant but talkative widow by the name of Ida Toney. Her deceased husband had been killed in a mine accident, leaving her with three small children to support, which she did by taking in boarders and by doing washing. She did, however, own a house on ground several hundred yards from the Vindicator mine property. Prior to this marriage, Orchard had been living in a cabin with Dan Kelly at Independence, but he was too much a loner to form a close friendship. He welcomed a move.[27]

The couple were married in June, and Orchard, carried away by his marital bliss, purchased furniture and did some minor repair work around the house.[28] For a time they were like any other married couple. Marriage seemed to be just the thing for both of them, although Orchard never told Ida of his former wife and child. Since they were in Canada, a thousand miles away, what difference did it make? He did show Ida some pictures of his wife and daughter, Olive, whom he passed off as his sister and her daughter. He told Mrs. Toney little about his early life except to say that he was born in New York State and later ran a cheese factory near Detroit, Michigan.

The third particular incident had to do with Orchard's employment in the Vindicator Mine. While he was working there, he started "high-grading" or, more simply, stealing high grade ore. The unusually rich ore might contain two, three, or even twenty dollars worth of gold per pound. The ore was taken from the mines in lunch buckets, shoe tops, secret pockets, or by some other imaginative means. Orchard started high-grading one day when he and his partner were working a slope alone. They hid the stolen ore between their underclothes and pant legs, where the pants were tucked into their shoes. Orchard later claimed to have taken out over fifty pounds of ore in one night in this fashion.

Mine owners estimated their losses due to high-grading alone ran over a million dollars a year. So common was stolen ore that it even turned up in the church collection plate. Hard to believe was the fact that no law existed in the district to prevent a miner for stealing it or an assayer from buying the stolen ore. The only grounds for which the "high-grader" could be prosecuted was trespassing, a common misdemeanor.

There were four assay shops in the little town of Independence, and the producing mines each had its own assayer. Most of the independent assay shops existed for the single purpose of dealing in stolen ore, usually paying the "high-grader" half the official price of gold. Occasionally one of the small assay offices would be blown up, an act that was generally accredited to the mine owners. Finding no legal way to stop high-grading, mine owners tried to eliminate the assayers so that the miners would have no outlet for their ore. When Bill Haywood became a power in the Western Federation of Miners, he further compounded the issue by encouraging the theft of ore, maintaining that the nation's mineral wealth rightly belonged to the men who produced it.

Mine owners tried various ways to stop the practice, such as installing "change rooms" where miners were compelled to change clothes when they were coming on shift and going off shift. But this did not check the stealing. Many insist that high-grading never really ended until the district ran out of the rich ore.

In the month of August, 1903, a strike was called by the Western Federation of Miners in Colorado City against the Standard Mill. According to Charles Moyer, president of the union, the reason was "the discharge of men from the mill for being members of organized labor." Union members in Cripple Creek were called out to cut off the ore shipment to the Standard Mill and force a settlement. Moyer publically testified that the members of the Cripple Creek district had no grievance against their employers, but were responding to the "call" to force the mine owners of Cripple Creek to bring pressure on the Standard Mill to submit to Federation demands.

The strike had been called by the executive committee of the Federation without taking a vote of the union involved. The committee had been given the power by the delegates to the annual Federation convention held in Denver in the spring of 1903. Officially, for publication, the union leaders of the Cripple Creek district called the strike, but in actuality it had been instigated through the executive committee of the Federation itself. And the plan of the Federation had been to call men out all over the state and completely tie up the mining industry. In this, they met success in some districts and failure in others.

On August 11, one day after the miners were called out, thirty-five hundred men were on strike in the Cripple Creek district. A strike committee was appointed and circulars soliciting aid for the strikers were sent all over the country. When the strike had first been called, it was conducted on an orderly basis. After the first two weeks, however, the strain on the mining community was felt. Business slumped, idle men agitated, and some of the once-healed wounds of 1894 bled again. Minor problems became major ones, and as tensions mounted, both union and non-union people armed themselves.

Differences between union men and non-union employees had been long and bitter. The Federation had learned from the Coeur d'Alene riots, subsequently adopting the permit system. Notices had been posted in the Cripple Creek district which read:

> Hence take notice that on and after September 15, 1901, anyone working in the mines, mills, and power plant of the Cripple Creek district who is unable to produce a card of membership in the Western Federation of Miners will be regarded as an enemy to himself and to the community at large and will be treated as such.

The notices were another attempt to reduce wages throughout the district. A second notice was distributed which pointed out that the fifteenth of September was near and the "time of grace" had just about expired. Now, all those who were not for the union must be considered against it — there could be no middle ground.

If a union man were brought before a justice of the peace who had non-union sympathies, he would be dealt with to the limit of the law. On the other hand, a non-union man brought before a justice of the peace who was a union sympathizer would be fined the limit.[29] In any case, the real hostilities of the district began when a non-union man was brought before a local justice of the peace, charged with carrying a concealed weapon, and was then released with an extremely light fine. In retaliation, union men caught up with the justice of the peace and beat him severely. Further incidents followed, and worried mine owners appealed to the county sheriff to call the governor for troops, but the sheriff refused, insisting that he could handle the situation.[30]

In September, leading citizens of the district petitioned Governor James H. Peabody for help. He responded by sending an investigating committee into the region to make an official report. As a result of this action, Peabody declared Cripple Creek to be in a state of insurrection because local authorities were unable to cope with the situation. State militia troops under the command of State Adjutant-General Sherman Bell soon arrived to take charge.

A few days after troops entered the district, Orchard and John Neville traveled to Denver for a spree and then on to Routt county for small-game hunting. Money was not a problem since Orchard had about two hundred and fifty dollars in savings. They were gone nearly a month. Orchard had thought the strike would be settled in that length of time, and had he known otherwise, quite possibly he might not have returned at all. Having gone through a strike with all the intensity of the Coeur d'Alene melee, Orchard was not eager to get mixed up in another.

The Cripple Creek strike, Orchard soon learned, was far from over; if anything, it was becoming more intense. The militia had arrested a group of Federation leaders including Davis, Easterly, Thomas Foster, and Sherman Parker, and had established a "bull pen" at Goldfield. Legal action was introduced on behalf of the prisoners when the judge of the district court issued a writ of habeas corpus and compelled the military to bring the union leaders into court and show cause for holding them.

The trials were held at Cripple Creek from September 21st to 24th. The court's decision held that a military commander ordered into state service by the governor had the power to make arrests, but he must immediately turn the arrested persons over to civil authorities. He ordered the accused men released. In defiance the military command announced that as an agent of the governor it had the right to retain the prisoners.

Federation leaders were delighted with this impasse and hoped the judge would direct the sheriff to enforce the order. Not only would a good fight probably result, but also it would look as if the military were persecuting the prisoners simply because they belonged to the Western Federation of Miners. However, a timely telegram from Governor Peabody ordered the military to release the union men, and a confrontation was temporarily avoided.

Orchard expressed the feelings of a great many miners at this time when he wrote:

> I attended the union meetings and felt more in sympathy with them, as I, like every one else, thought they were persecuting these men because they were active union men,

> and I hated the militia more than I did the non-union men. But I hated them all, and I felt more bitter against them all the time.

During this period of inactivity, Orchard returned to high-grading in the Vindicator mine. He was no longer employed, he needed money, and he was bored because of inactivity. A detachment of the military troops was camped only a few hundred feet from the shaft which was used to descend into the mine, and their presence increased the excitment Orchard felt because of the risks he was taking:

> This 'high-grading' was no easy job, as we had to climb down an old man-way nine hundred feet, where the ladders were out in some places, and then go through old stopes and drifts two or three thousand feet, dig out our load, and pack it back. We would pack all the way from forty to eighty pounds.

During one of his forays underground, Orchard discovered a carload of dynamite in a cross-cut on the eighth level, a clear violation of mine safety. He was astonished and yet fascinated. What a man could do with a whole load of powder at his disposal! Some time later when he met Davis, the president of the Altman Union, Orchard mentioned the carload of explosives and jokingly suggested that if the union ever wanted to create a sensation, it could blow that up. Suddenly, the talk became serious, and Davis officially began querying how such an explosion could be set off. Several days later, he contacted Orchard and asked if he would be willing to enter the Vindicator and set up an explosion that would detonate when a non-union shift was at work.

Davis felt that such a violent event was imperative. A graphic demonstration of force would put the fear of God into the scabs who were arriving daily in the district, and Federation members would also be impressed by this show of force and would result in closed ranks and in stemming the flow of deserters. Solidarity was necessary at any cost, he felt.

Davis clinched his suggestion with an offer of two hundred

dollars. As usual, Orchard needed cash, so the offer was tempting. The only catch to the plan was that Davis did not have any ready money. He claimed they had had so many failures such as blowing up the power plant at Colorado Springs that he hated to ask for money until something sensational was actually done. Federation headquarters in Denver was his only source, an indication to even the lowest rank and file union members, such as Orchard, that involvement extended to the highest level of the Federation.

Orchard accepted the offer, and his new career was launched.

CHAPTER 7.

To think they would go out and get some one else to do an easy job like that, after I had taken chances down in the mine, and right under the very noses of the soldiers.
. . . Harry Orchard

Orchard quickly went to work. To help with the job, he asked a friend, Joe Schultz, who had high-graded with him. He and Schultz, a runt of a man with a heavy mustache and constantly sore eyes, had previously stolen two fifty pound boxes of dynamite from the underground carload and had sold them to independent mine leasers. Schultz, too, needed money and claimed he was justified in his actions because of the influx of scab labor to the mine.

The pair made their way into the mine just after midnight, timing their approach so they could move at random in the underground workings while the midnight shift went to the surface to eat. As they approached the eighth level station, which was lit by electric lights and supposedly deserted, they were taken by surprise by a cager still on duty. He assumed they were a couple of late miners since he had just sent up a load of men to the top.

"Hurry up, boys, this is the last cage," he snapped.

They instinctively backed away from the bright lights, plunging into the darkened tunnel as fast as they could, not daring to use their lights. The cager, confused by their actions, followed them. Both Orchard and Schultz had revolvers. They turned and fired, but missed, and the cager retreated in the opposite direction.

Panic-stricken, Orchard and Schultz tried to get out of the mine as soon as possible, not bothering to look further for the carload of powder. Schultz tried to invent a reason for being in the mine, if they were captured. The best he could come up with was a wild story of counting how many machines were running in the Vindicator. Orchard, in a flare of bravado, rattled on about shooting his way out if they encountered guards.

They managed to emerge from the shaft and leave the area unnoticed. The sight of the sky and the peaceful night, after the terrifying underground experience, was something Orchard never forgot. When he saw Davis. he told him of the encounter with the cager, and to impress the man, Orchard invented a story that after the shooting, he and Schultz had continued to the spot where they had last seen the carload of explosives, but it had been removed. Unknown to Orchard, this part of the story was true since the explosives had been moved to the powder magazine where it should have been stored in the first place.[31]

Since Orchard had failed to blow up the mine, he received no money. He went to Davis and Sherman Parker, who had replaced Easterly as secretary, to borrow money with which to pay his wife's taxes on mining property she owned. Both men hedged at making a loan. Parker told him the chances of getting any money were pretty slim. Now if he had killed Harrah, the mine cager, Orchard would have received all the money he needed, but alas, he had produced nothing.

Some days later Davis and Parker approached Orchard again about blowing up the Vindicator, or perhaps the Finley mine. They suggested that this time a powder charge be rigged near a mine shaft, so that when a cage loaded with non-union miners descended, the charge would go off and wipe them all out. Orchard pointed out that since the cage was often sent down empty, it was possible that only the cage structure itself would be destroyed. He suggested that they come up with a technique that would kill as many men as possible at one time.

It was Parker who contacted Orchard and offered him five hundred dollars if he would "fix something in either mine to kill some of them so as to scare the rest and make them quit, and keep our men from going back to work, and scare outside men from coming in there to work." Orchard decided to sneak into the mine after the night shift quit and rig up an explosive charge. But how would he set it off?

Federation leadership seemed to prefer the use of explosives whenever possible, since they liked the terrifying effect they produced. The horrifying use of powder would also be identified in the public mind as the kind of violent action that the Western Federation of Miners would resort to in order to achieve its ends.

Orchard sought the man who knew the most about explosives — Bill Easterly, an expert with powder work. The two began experimenting with explosives near an abandoned building on Bull Hill, between the Isabella mine and the Independence. Finally, they tied a string around four or five blasting caps and fastened them in a solid position. A six-shot revolver was then placed two or three inches from the caps and firmly mounted against the side of the building. A strong cord was then attached to the trigger, and when it was pulled, the revolver discharged and shot directly into the blasting caps. When it was tested, it worked to perfection, with only one drawback: the revolver was destroyed.

What if the revolver were fastened to a timber of the shaft at a level station? A wire could be attached from the trigger of the revolver to the station guard rail. When the guard rail was raised, it would put enough pressure on the trigger wire to fire the weapon, thus setting off the powder charge. In this way, the guard rail became the key to the death trap. It was simply a safety device that was raised as the men got out of the cage and then lowered to prevent anyone from falling into the shaft. Orchard's death trap could be installed on any station in the mine.

A sophisticated method of murder had evolved, the ingenuity of which would become Orchard's trademark. A fascination for complex destruction devices had been aroused, and the inventor was proud of his ingenuity. Europeans had used explosives for assassination purposes for some time, but Orchard's technique was a vast departure from the accepted European method of throwing a bomb and then risking being killed along with the victim. He had devised a way in which the victim would kill himself but allow the murderer to escape uninjured.

Orchard naturally wanted Easterly to assist in setting up the death trap, but he refused. Somewhat disgruntled, he later claimed:

> Easterly did not go with me, because none of these active labor leaders did anything themselves, if they could help it. They always managed to be in some conspicuous place when anything was likely to happen.

So Orchard then contacted his old buddy, Schultz, and asked if he wanted another go at it. Schultz emphatically said "no." Orchard then carried his troubles to Parker, who came up with the name of one Billy Aikman, a man who wasn't afraid to spill a little blood. Orchard put the proposition of murder to Aikman, and he accepted.[32]

Perhaps the reason Aikman so readily accepted such an adventure was that in killing a person by using a booby trap, he would feel detached from the actual murder itself. He did not have to face his victim eyeball to eyeball, see the expression of fear and horror on his face, or watch the body as life left it. He did not have to face the option of giving his victim a last minute reprieve. The booby trap would set him apart from all this. It would become death at a distance.

Orchard began to set things up for the big kill. He obtained powder from Joe Craig, his wife's brother, who leased a small mine. He also got some powder from his friend Billy Gaffney, and since he thought a look-out man might be useful, he finally persuaded Gaffney to join him. After Orchard had collected his equipment, he and Aikman met at Gaffney's house, which was also situated near the Vindicator property, and grimly began assembling the materials of death and destruction.

A little after two-thirty on a November morning in 1903, after the night crew left, the three motley conspirators carried nearly fifty pounds of dynamite, a box of blasting caps, and a revolver Orchard had obtained, down the mine shaft. Gaffney was left a short distance down the shaft as a look-out while Aikman and Orchard continued further underground and eventually reached the sixth level. Actually the two men thought they were on the seventh level, despite the fact that there was an abundance of dust in the tunnel, which normally would have indicated to them that it was an unused level. But since they were under abnormal pressure at the time, they did not notice the collection of dust and assumed they were on the seventh level as planned.

On reaching the level station, the death trap was hurriedly put in place, and they left as soon as possible. Nearing the surface, they searched for Gaffney, but he was not to be found. Where in the hell was the son-of-a-bitch? Had something gone wrong? There was no place to go but out of the mine. As they left the mine shaft, Orchard picked up a bottle of turpentine that they had secreted and poured it on the ground to cover their tracks, in case dogs might be used to hunt them. Later they found Gaffney and demanded a damn good excuse for his absence in the mine shaft. Gaffney explained that he had begun coughing while he was underground, and rather than give anything away, he decided to leave. His answer seemed reasonable enough and was accepted.

The next morning, Davis woke Orchard up out of a sound sleep and wanted to know if everything was in place. He was assured that it was, but still there had been no explosion or excitement at the mine. Not completely satisfied, Davis left the house and Orchard went back to sleep. Rising about noon, Orchard ate and then went over to Billy Aikman's house and asked if he had heard anything; he had not. Apprehensively, the two walked over to the Vindicator property where they found everything normal and in operation. They decided that the bomb was a failure, and it was a week before they learned differently.

During the time lapse between setting up the bomb and the actual explosion, Orchard again tried to obtain money from Parker and Davis, without success. During their conversation, however, Orchard learned that Federation men had tried to derail the Florence and Cripple Creek train that was carrying non-union miners, but they had failed as they had broken their tools before the job could be completed. Undaunted, they intended to try again at a point between Victor and the town of Cripple Creek. Many of the non-union miners lived in those particular communities and commuted by train to the mines on Bull Hill, since it was safer to live there than in the more radical union areas.

It was the wrong time and the wrong thing to mention to Orchard. He felt intensely hurt, jealous, and angry.

> To think they would go out and get some one else to do an easy job like that, after I had taken chances down in the mine and right under the very noses of the soldiers.

Orchard was enraged enough to blow the whistle on the

Federation and the railroad authorities about the attempted derailment. The next afternoon found the frustrated dynamiter in Cripple Creek, tryng to locate someone in authority. Quite by accident, he ran into an acquaintance, John Jones, a conductor on the line, and he referred Orchard to D.C. Scott, a detective for the Florence and Cripple Creek railroad. Scott's office was over the depot at Cripple Creek. Finding him in, Orchard introduced himself as a fellow Mason, and with little preliminary talk launched into some details on the attempted derailment. He told Scott that "some of the Bull Hill boys" were going to wreck a train and succeeded in convincing the investigator that he did possess facts and knowledge concerning the case. The discussion terminated with Orchard agreeing to see Scott if he found any more information.

On November 21, 1903, the superintendent, Charles McCormick, and his shift boss, Mel Beck of the Vindicator mine, were making an inspection trip to the unused sixth level, to check the possibility of reopening the level. As the cage stopped at the level station, one of the men reached forward and raised the guard rail, and suddenly the two men were blown to eternity. The death trap had worked perfectly. The wire wrapped around one end of the guard rail and attached to the trigger of the revolver, firing it as the safety rail was moved upward. The shot hit the box of blasting caps and set off the buried dynamite, leaving torn, lifeless bodies and the heavy, acrid smoke of dynamite. Harry Orchard had graduated to the criminal status of murderer.

One of the first to reach the murder scene was Major H.A. Naylor of the Colorado militia, who arrived at the mine at approximately twelve o'clock. He was delayed a short time while a new cage was put on the cable, since the blast had ripped the original cage to shreds. At the scene of the explosion, Naylor observed the wrecked station and then choked violently on the fumes of spent powder. McCormick was lying on the north side of the tunnel and Beck, on the south. Both men were beyond help.[33]

Major Naylor found wire still attached to the shattered guard rail and to parts of the revolver. Exactly what had taken place on that doomed mine level remained in doubt until Orchard's confession clarified the mystery. Then, and only then, was the crime satisfactorily explained.[34] The bizarre element of the murder was that Orchard knew both men and considered them "swell fellows." He had not wanted to kill those two men. It was all a mistake. "I thought it would kill a cage-load of non-union men, as the men always went down first going on shift," he later said.

Even the nervousness he felt and the second thoughts that haunted him because of his actions could not keep him from viewing the broken and bloody bodies at the coroner's office. That night he saw Davis and Parker at the miners' union hall at Victor. They boosted him out of his momentary depression by praising his act of murder. Orchard responded to praise. Assured that money would be forthcoming, he went home to bed and found that a murderer can sleep just as soundly as any other man. And that was just the beginning for Harry Orchard.

CHAPTER 8.

I introduced myself, told them who I was, and they said they knew me by reputation . . .
. . . Harry Orchard

On Monday, the 23rd, detective D.C. Scott summoned Orchard to Cripple Creek. Harry was extremely wary. The Vindicator mine explosion had placed him in an entirely different category since his first meeting with the detective. He was now a murderer, and the last thing he wanted right now was to talk to a detective, any detective. Orchard realized he had to put on a front, and a good one, if he didn't want to arouse suspicion concerning his role in the labor trouble.

When he arrived at Scott's office, greetings were exchanged and Orchard took the chair offered him. After several minutes of small talk, Scott asked him if he had any objections to speaking to K.C. Sterling, a detective for the Mine Owners Association. Harry felt cornered — fear knotted his belly, and he fought back a sudden impluse to vomit. Outwardly, however, he remained calm and replied to the question in the negative.

Sterling arrived, shook hands, and immediately launched into the details surrounding the Vindicator explosion. Orchard maintained his composure and suggested a picture of complete innocence as he denied having any special knowledge. He even advanced a personal opinion that the explosion was possibly accidental. Abruptly changing the subject, Sterling asked point blank who had told him about the attempted train derailment. To protect himself from being more involved, Orchard refused to answer. Not wishing to push the matter farther and thus alienate what appeared to be a prime informant, the detective didn't press the matter.

Ironically, it was the murderer's association with these two detectives that saved him at a later date from being driven out of the district together with other hoodlums. His position as a possible informant prompted Sterling to add his name to a list of mine company spies, and he was ironically protected. Orchard never realized how this odd set of circumstances benefited him.

After that, Harry Orchard had several meetings with the

detectives. Not wishing to expose himself and his connection with the Federation, he always appeared when he was summoned, but he never released information that would jeopardize him or his cohort. Scott gave Orchard twenty dollars and offered to pay him a hundred dollars monthly to work for him, but Orchard refused, claiming his sympathy lay with the union cause. No doubt Orchard found a certain amount of thrill attached to his arrangement with the detective. Playing the game from both sides of the field fed the murderer's ego. It was his opportunity to be someone important.

In the meantime, four men — Sherman Parker, William Davis, Thomas Foster, and H.H. McKinney — were arrested for the attempted train derailment.[35] The mining district was shaken when it was rumored that McKinney had confessed and had implicated the others. Now Orchard really began to sweat. Had Parker told McKinney and the others about the Vindicator explosion? If so, had McKinney already spilled his guts about that? The more the killer thought about it, the more he drank; and the more he drank, the more he thought about it. He had to find out for certain, but how? Why not ask Parker himself! The fact that Parker was in jail awaiting trial was only a minor obstacle.

With all the confidence and determination of a carnival medicine man, Orchard went directly to Scott and asked him to fix it so he could see the prisoners.

"What for?" questioned the detective.

"Just want to say hello to the boys and give them a bottle of liquor and some cigars," replied the nonchalant Orchard.

Scott granted the request by telephoning the sheriff and asking admittance for Orchard. It appears that he considered Orchard trustworthy enough to grant such a request but also must have hoped that some scrap of new information might come from the meeting. As it turned out, both men were disappointed. Orchard was not given the opportunity to speak privately to Parker or Davis because of the ever-present guard, and Scott discovered that the visit turned out to be just what Orchard had predicted — strictly a social call.

Undaunted, Scott knew that Federation attorneys had seen McKinney in jail and had extracted a statement, the importance of which could affect the case against the union men. So in December, 1903, he proposed to Orchard that they take a trip to Denver to see if anything could be learned about McKinney at Federation headquarters. Orchard did not need much persuasion, because Bill Easterly was reported to be in Denver and was acquainted with both Moyer and Haywood. Easterly could get the money still due Orchard for the Vindicator job. He picked up a railroad pass and, with the help of some expense money from Scott, the two left for Denver the next afternoon. Once there, they were careful not to be seen in each other's company, even to the point of staying in different hotels.

The following morning Orchard hurried to the Western Federation of Miners' headquarters in the Mining Exchange Building on Eighteenth Street. Boldly he requested an immediate audience with the union leaders, William Haywood and Charles Moyer. Orchard recorded the meeting, "I introduced myself, told them who I was, and they said they knew me by reputation!" Guardedly, the men acknowledged the murderer, but told him that Easterly was in Pueblo and would be gone a day or two. They made it clear that his presence was necessary for identification purposes before they could discuss business and financial arrangements. Moyer gave Orchard twenty dollars to tide him over until Easterly's return.

Orchard then met Scott at the latter's hotel and told him he had learned nothing. Scott accepted the news stoically, and since he had to return to Cripple Creek, he prevailed upon Orchard to remain in Denver until Easterly returned. He gave Orchard more money, bringing the total amount spent on the Denver venture to about forty dollars. The killer accepted the money and felt exhilerated with his newly-found manipulative powers of drawing currency from both Moyer and Scott. And the thrill of the secrecy bolstered his ego another notch.

When Easterly returned, the men met, and, according to Orchard, Haywood and Moyer complimented him on the Vindicator job. Haywood carried the discussion one step further, saying they "got two good ones, and such were hard to get; a few like them and we would have everything our own way." As one of his future projects, Haywood pointed out that none of the "tin soldiers" had been hurt. It was now their turn.

When the question of money was brought up, Orchard was told not to accept too much, as he could get more any time he needed it. The killer settled on three hundred, which Haywood delivered, although not at once. Haywood also stressed the hazard of displaying it. A payment, a little at a time and spread around, was the safest way. This financial arrangement seemed to be a standard method of operation for Federation officials and was quite successful.

Needless to say, both Moyer and Haywood always denied this particular meeting by maintaining they had never met Orchard until a later time. But it was common knowledge that Harry Orchard had money during the entire length of the strike, a period of time when money was at a premium, and he had not been visibly employed since the beginning of the strike. The union claimed that Orchard's support came from the miners' strike fund, that he supposedly received benefits from the Federation just the same as any other union miner. At any rate, unemployment did not prevent Orchard from enjoying his beer, his whiskey, his gambling, and his whoring. One noticeable feature was that his clothing improved, but when Orchard had cash, he always dressed well.

His wife Ida was later to remark that her husband looked and felt very important after he returned from Denver. Something had happened there. He was praised, made to feel important, and he thrived on this new-found prestige. After all, he was now a professional assassin for one of the most powerful labor unions in the United States. Harry Orchard had at last arrived.

Orchard paid Billy Aikman, his partner in the Vindicator explosion, one hundred dollars, he skillfully neglecting to mention the amount of money he had received for the job. Billy Gaffney, the faint-hearted lookout man, was given a dollar or two periodically. Orchard was afraid to give Gaffney too much money because of his heavy drinking and his inclination to "shoot-off" his mouth. Although Orchard was aware that liquor and secrecy didn't mix well, it did not deter his own drinking habits.

During the first part of January, 1904, Harry Orchard struck out on his own. Dynamiting had brought out Orchard's real skills and abilities. He was original in his use of explosives for destructive purposes, an innovator in the forceful use of powder. Most important, he knew the necessity of his being away from the scene of the disaster.

Orchard's next scheme was an attempt to blow up the Vindicator's huge boiler system. The plan originally had been to throw explosives into the mine's coal bin, but it occurred to him that they would be readily detected when the coal was being moved. Thus he became determined to disguise the dynamite in such a manner that it could not be distinguished.

Tying up with a boozing buddy, Owey Barnes, the two men stole ten pounds of dynamite and blasting caps from a leaser by the name of Charlie Perkins and moved the stolen articles to Orchard's house where they hid them in some old out-buildings. Since Orchard's step-sons were always around the place, the two men transferred the stolen explosives to Barnes' cabin the following morning.

The two conspirators procured a tin can into which they placed the dynamite and caps. They then packed asphaltum around it so it seemed like a solid mass, and the deadly black bomb was ready. Two bombs were fashioned in a little over an hour, then Orchard took one of them to Billy Aikman's saloon, and Aikman gave it to an old man by the name of Dempsey. Dempsey had taken an active part in the union troubles of 1894 and had shot a couple of deputies during that period. Since he was an older man, he would have no problem getting past the soldiers. Dempsey promised to throw the bomb into the Vindicator's coal bunkers so that the boilers would be blown apart.

Later that same night, Dempsey telephoned Aikman's saloon and talked personally to Orchard. The man was drunk. Orchard heard him say he had delivered the goods, then cut him off. He was angered to think Aikman had hired a man like that to do the job, the drunk son-of-a-bitch! Subsequently, Dempsey maintained that he had thrown the bomb into the coal bunker, but nothing came of it. Was it in some way defective, or had Dempsey never accomplished the task? The answer remained a mystery for years.

What actually happened was that Dempsey did place the bomb in the coal bin. Asphaltum, however, except under proper manipulation, does not become adhesive, but will instead break up like a clod of dry soil. The black bomb had been shattered when it was thrown into the bin; the sticks of dynamite were exposed, and as chance would have it, a fireman saw them. He handed the bomb to the powder man in the mine, W.G. Ramsey, who further examined the explosive device and found it contained ten sticks of dynamite. Orchard gave the remaining bomb to Easterly several months later, and that was the end of that plan.

The Colorado State Federation of Labor called a convention for organized labor to convene on January 11, 1904. Several important items were to be considered. One was to discuss ways and means of defeating Governor Peabody for reelection, and another was to raise funds for the embattled Cripple Creek district. Acknowledging the convention, the Western Federation of Miners sent word to its district leaders where strikes were being carried on to send its best men to Denver for the convention, but not to include the strike leaders as they were needed on the district level to hold the men in line. From the Altman Union No. 19, considered at that time the most radical union in Colorado, came two delegates — Paddy Maliney and Harry Orchard.

The selection of Orchard as a delegate was testimony to his standing among local union officials. He was never noted as being a working, rank and file union man. For the most part, he was a non-participant who attended very few meetings during the years he was a Federation member. Nonetheless, the two Altman delegates appeared to be close friends. They roomed together, drank together, and chased hustlers together. Both spoke along radical lines when the leaders were considering solutions to problems. It was also evident that both men were advised by Moyer on each important question as it arose.

The most important committee of the convention was the Ways and Means Committee which was in charge of campaign funds and was representative of the entire labor force in Colorado. A dispute arose concerning the delegates who should represent the Western Federation of Miners on this committee. The president of the convention, J.C. Sullivan, favored Maliney of Altman, while Moyer, president of the Federation opposed it.

The impasse was finally settled by leaving the choice of the two Federation representatives on the Ways and Means Committee to the president of that union, Moyer. He appointed himself and the unknown Harry Orchard.

When the convention adjourned on January 14, Orchard had revealed very distinct personality traits. He was described by one delegate as a rather weak man who could be dominated by others, a description with which even Harry Orchard agreed. The convention, however, noted quite a closeness between Moyer and Orchard, and the next time the State Federation of Labor heard of Orchard was at the convention held in Pueblo, Colorado, in September, 1904. President J.C. Sullivan reported that he had been notified that Harry Orchard of Teller County had gone somewhere else to seek employment. Sullivan recommended that another delegate be chosen to represent the Western Federation of Miners on the Ways and Means Committee, since Orchard was wanted for murder at the time.

Orchard had talked with Haywood at the convention and was emphatically instructed not to pull anything in the district until the trials of the attempted train wreckers were concluded. Don't give the prosecution and the public any more ammunition to use against the defendants, he was warned. Orchard understood the implications.

It was at this time when Harry Orchard, dynamiter, met George Pettibone, former Coeur d'Alene resident and dynamiter.[36] Pettibone was not a Federation member *per se*, but maintained an honorary status. His deep friendship with both Moyer and Haywood gave him the privilege of having both as confidants. Certainly the two leaders heavily relied on him for assistance and information. At the convention Pettibone acted as doorkeeper, a convenient position for an opportunistic individual to watch the pulse and the direction of the statewide delegates. Moyer had previously discussed with Orchard a certain deadly chemical mixture Pettibone was familiar with, something called Grecian Fire or Hell-Fire.

Late one night Orchard went to see Pettibone at his store — or to see the "devil," as Moyer humorously called him. In the basement of Pettibone's store, a most unusual experiment took place. The older man spoke excitedly as he demonstrated the mixing of chloride of potash and sugar and then set it on fire with sulphuric acid. This could be used to set off detonating caps, causing a big explosion. Orchard immediately grasped the significance of such a preparation and considered the possibilities.

This chemical mixture, the so-called Hell-Fire or Grecian Fire was something else. This dangerous concoction was composed of stick phosphorus, bisulphide of carbon, alcohol, spirits of turpentine, and benzine.[37] When combined properly, bottled, and thrown, it would burst into flames on contact. The chemical could also be blended so that the time of firing could be determined.

In handling Hell-Fire, one had to be particularly careful. If just a drop of the mixture accidently spilled on clothing or some part of the body, it would begin to burn. Once Pettibone had dripped some on his shoes, and in attempting to clean them, they began to smoulder. Pettibone once took Marion Moore, a member of the Federation Executive Board, out in the country to watch him demonstrate the concoction, and a little splashed on his coat. He soaked the coat in water, but when it was dry, it began to burn again.

It was suggested that Orchard take some of the Hell-Fire back to Cripple Creek with him. A good terror tactic would be to throw it down a mine shaft to set it on fire. Better yet, it could be thrown on non-union workers while they were riding on the work train. Oh, it had innumerable uses, all of them extremely lethal.

Pettibone eagerly secured enough of the necessary ingredients to make several gallons of the Hell-Fire. Orchard was especially skeptical about carrying it into an area which was saturated with military troops. Nonetheless, carrying the valise full of chemicals in his arms and keeping it close to him the entire trip, he boarded the train for the mining district. Once home, he put the bottles in a shed on his property and told his step-sons not to touch them. The inquisitive boys wanted to know what they were. Orchard told them they were used to kill bears. After this, he took further precautions and buried them in his yard.[38]

Orchard's convention trip to Denver had been a big success. The murderer had been placed on the most influential committee in the state labor organization; he had met the "devil" Pettibone himself, and he had gotten some new ideas and slants on creating sheer holocaust which was now his specialty. More significant was his enthusiasm to "tear something loose" at the first opportunity. All he needed was the opportunity to let loose. For a man of Orchard's dedication, that opportunity was not too far away. It was a time of violent loyalties and fierce partisan struggles. It was a time when anything could — and frequently did — happen. And Harry Orchard was determined to be an important part of the action.

CHAPTER 9.

He seemed a very energetic worker.
. . . Charles Moyer

Several days after Orchard had returned from his Denver trip, Pettibone arrived in the mining district to give his promising protege further important instructions on the deadly Hell-Fire. Welcomed, Pettibone was introduced to Ida and the boys as Mr. Morgan. When the two men were alone, the merchant handed the killer three glass bottles containing alcohol, benzine, and turpentine.

Gleefully, Pettibone demonstrated his technique. He put phosphorus in bisulphide of carbon and added benzine until the mixture began to smoke. Then he added alcohol and turpentine. The bottle was sealed with a glass stopper. The concoction had one very distinct characteristic — an extremely offensive odor. On the whole, the meeting was most enlightening to the murderer. But then, he always learned many new things from George Pettibone.

In district court during the month of February, 1904, the attempted railroad wrecking case against union men Parker, Foster, and Davis reached the trial stage. The men were charged with conspiracy to commit murder, and the prosecution seemed intent on crucifying them. Davis, however, was released after the opening of the trials for lack of sufficient evidence. Now, the prosecution's case rested on the testimony of H.H. McKinney, a one time deputy sheriff of Goldfield. According to his statement, Parker had offered him several hundred dollars to derail the train carrying non-union laborers. Foster, according to McKinney, had assisted him in the first unsuccessful attempt, and a Charles Beckman had helped him the second time.

Charles Beckman testified that he had become a member of the Western Federation of Miners as a spy for the Mine Owners Association. Detectives Sterling and Scott further testified that they had received direct information from Beckman and had observed the second train wrecking attempt. Mrs. McKinney was brought to the stand and corroborated her husband's testimony on some points.

When it was their turn, the defense came on strong, attempting to show that McKinney had been bribed and frightened into making a perjured confession to convict the accused union leaders and throw the union movement itself into disrepute.[39] The defense then maneuvered its case to show that McKinney had been given highly unusual privileges while he was in jail — such as being sent out for meals and given time to spend with his wife away from the confines of jail.

For emphasis the defense produced a butcher employed in the union shop at Victor who testified that he had walked along the tracks on the day in question and had seen two men tampering with the rails. He shocked the court by identifying the two men as none other than detectives Scott and Sterling! And numerous witnesses came forward to swear that Beckman had tried to persuade them to commit a variety of crimes to win the strike.

Meanwhile, the jury listened intently. It was composed of men who were living outside the mining district. None were employed in the mines or associated with the Federation. Finally, E.A. Eisworth, jury foreman, read the verdict. It was a unanimous agreement, reached after a single ballot and within five minutes after they had retired from the court room. It was read to a hushed audience:

NOT GUILTY

The case of the people of Colorado vs. the Federation of Miners was over.

Orchard remarked that Federation men had been ready to perjure themselves on any point the defense desired. He had fully expected to be called as a witness and would willingly have given false testimony, but then he could not very well have done otherwise without seriously implicating himself.

At one point duirng the trial, Moyer became overly worried and concerned. McKinney was giving testimony about Parker's telling him about the strange liquid that would burn like fire:

> . . . he also said he had experimented with it and had tried a bottle full of it, but it didn't break when he threw it against something until it struck the ground, and he said the flames jumped up ten feet high.

For some unexplained reason, the prosecution did not catch the significance of the statement and so failed to pursue the subject. After that, Moyer breathed easier. But Orchard was still gripped with fear that his involvement in the Vindicator explosion might slip out into testimony, particularly if one of the defendants had ever mentioned that fact to McKinney. Luck was with him as they apparently kept the knowledge and their intentions to themselves.

Moyer regularly attended the trial. He assisted the defense and saw Orchard nearly every day. He was impressed with the murderer, even observing that, "He seemed a very energetic worker." For his part, Orchard made use of his new found influence with the labor leader and informed Moyer he would deal directly only with headquarters on assignments. He suspected that Davis and Parker were holding out on money that was rightfully his. His suspicion was shared by Steve Adams. So before he left Cripple Creek, Moyer slipped his favorite killer a hundred and fifty dollars and promised he would be given another mission in the near future.

Soon after the trial, Orchard went to Denver at the summons of Moyer and Haywood. Moyer had a special job for the killer: to act as bodyguard for the Federation president when he visited unions at Ouray, Telluride, and other camps in the southern part of Colorado. Union miners had been deported from Telluride several days before, and Moyer wanted to see what could be done for them. Feeling was running strong against the Federation, and Moyer feared for his own personal safety. As an added precaution, he decided a little extra protection would probably be useful, and Harry Orchard was just the man for the job.

It was at this time that Haywood claimed to have met Harry Orchard for the first time. The meeting, he said, took place at Federation headquarters as Moyer and Orchard were getting ready to leave for Ouray. Haywood testified:

> I asked Orchard if he had a gun with him. He had, and he pulled a six-shooter a foot and a half long out of his pants. I said to him: 'That's not a very handy gun; you'll have to pull your pants off to get it in action.' He looked as though

he did not like criticism, but he didn't say anything.

Actually, Orchard was better armed than this. He was carrying two pistols, one that he had borrowed from Joe Craig and one that he had received from Joe Schultz. Moyer had a pistol and two sawed-off shot guns that could be carried in their luggage. They were prepared for anything.

Telluride was an important silver and gold mining camp in the San Juan district of southern Colorado. It, too, had a bloody history of labor warfare. The president of the local miners union of the Western Federation of Miners was a wild radical by the name of Vincent St. John who had been elected to that office in February, 1901. He was no stranger to violence; he courted it and slept with it like a mistress.

The Smuggler-Union was one of the prominent mines in the district. The manager, Arthur Collins, had instituted a controversial "contract" or "piece system" of mining, under which a miner was given a contract to break ore at so much money per fathom. When the system had been first introduced, miners were required to break only the ore, but later it became mandatory that the ore be broken down to a suitable size and then shoveled down the mine chutes. Collins' position was that the contract system was optional with his workers. They could work by the day if they wished, but the system could not be altered. And so a strike was declared on May 1, 1901, by the Federation against the Smuggler-Union properties to force the abolition of the contract system.

The Smuggler-Union mine finally reopened with non-union labor, and it operated successfully, too successfully, until a large group of heavily armed union miners descended on the place. At daybreak on July 3, Federation men took positions near the mine headquarters. When the night shift came off and the morning shift was about to go on, shots were fired into the buildings. The unarmed workers were taken by complete surprise; bullets ricocheted dangerously, and the smell of death was inside, along with the odor of gunpowder, outside. When the fighting was over, there was one dead union man, killed accidently by his own men, and two non-union miners were dead. Among the wounded was Charles Becker, mine superintendent.

The non-union miners were lined up and forced to march over the rugged mountains and out of the district. In the process, they were beaten and harassed. A strange document was also drawn up, reminiscent of the 1892 agreement of the Coeur d'Alene labor dispute. Here is a portion of that agreement:

> It is hereby agreed between the Miners' Union, by V. St. John, president, and the Smuggler-Union Mining Company, by Edgar A. Collins, assistant manager, that all work on said mine shall cease for a period of three days, ending Friday, and that said Miners' Union will refrain from violence as to person and property for the same period.The said Smuggler-Union Mine may employ four guards during the period.

Perhaps never before had the temperament of the extreme elements of the Federation been more apparent. The statement that the union would refrain from violence to persons and to property for a specified time period was in itself an incriminating indictment of the violent aspects of the movement.

A settlement was reached, however, and the contract method of mining was discontinued. The owners were permitted to take charge of their property. The conflict did not end, though, but continued with regularity. Murders and beatings became part of Telluride's existence, the climax of which came on November 19, 1902, when Arthur Collins, manager of the Smuggler-Union Mine, was killed by a shotgun blast fired by Steve Adams while Collins was playing cards with friends.[40]

Another strike, this one demanding an eight-hour day, was called by the Federation on September 1, 1903, in support of the mill and smelter workers in that section. Armed union men were stationed at some of the mines, which remained idle for several weeks until Governor Peabody sent the militia to protect the non-union labor which the companies had brought in to work. Martial law was proclaimed, and the mines were once again opened. But to keep them operating, the militia had to be on duty all winter and into the spring. When the militia started deporting the hated Federation men from the district, Moyer felt he had to go to Telluride to determine how the union could best help its members.

Moyer and his number one henchman, Harry Orchard, headed for Montrose where John Murphy, a Federation attorney, was trying to obtain a court injunction against the militia, to compel them to let the union miners return to their homes. Arriving in Montrose, they were met by Murphy, who had obtained the court injunction. On March 24, 1904, they went to Ouray where most of the deported men were staying. The next day Moyer sent a telegram to Governor Peabody informing him of the injunction and asking if he intended to honor it. The Governor's reply was that all law-abiding citizens would be protected. The following morning Moyer sent a few miners over to Telluride to test the situation. They were intercepted some distance from town and threatened with death if they attempted to enter. Sherman Bell, the Colorado adjutant general, was in Telluride in personal command, and he was determined to ignore the court order.

When the union miners returned to Ouray, Moyer was furious. The only advice he could give the despondent men was to take the law into their own hands, go back to Telluride, and clean out the town. How to do this was the problem! Since Telluride, like most mining camps, was located in a canyon with high mountains on each side, they decided that beer kegs might be loaded with dynamite and rolled down the mountainside into the town.

Moyer himself advanced a plan that would have appealed to any madman intent on murder. The scheme called for dumping cyanide of potassium into the town's water reservoir. Getting the deadly poison would be no problem since cyanide of potassium was available in mills where the cyanide process was used. Fortunately for the citizens of Telluride, Moyer was arrested before his proposed plan could be carried out.

One plan, however, that was almost carried through, just before Moyer was taken out of commission, involved the relocation of the deported miners. For financial expediency, it was decided that the Federation should lease some unused mines on Red Mountain, about halfway between Ouray and Silverton and fewer than twenty miles from Telluride. There were some large buildings, boarding and lodging houses. One large dividend would be that the men would be concentrated in one area. Vincent St. John, the fireball president of the Telluride union, was to be sent to the encampment when it was established, to organize the miners into a paramilitary unit for sustained combat operations. Only a matter of hours kept the plan from being set in motion.

On March 26, the very morning that the Red Mountain plan was to be completed, Moyer and Orchard, who were sharing the same room, were awakened by a constant and determined knocking on the door. When Moyer answered the door, he was confronted by the local sheriff who immediately placed him under arrest. Orchard, in the role of the ever-alert bodyguard, was taken by surprise before he could defend his employer.

Moyer was arrested upon the request of Sheriff Rutan of San Miguel County, of which Telluride was the county seat, and was taken to Victorburg which was half-way to Telluride. The officer who arrested him had no warrant, and Moyer refused to budge until a warrant was served. Sheriff Rutan then served a warrant charging desecration of the American flag at Telluride in September, 1903.

The charge of flag desecration grew out of a Federation poster which had been designed by Haywood and O'Neil. At the top of the poster was the caption, "Is Colorado in America?" A picture of the American flag was the focal point of the poster.[41] On each stripe of the flag were the following actions taken against the state of Colorado:

Martial Law Declared in Colorado.
Habeas Corpus Suspended in Colorado.
Free Press Throttled in Colorado.
Bull-pens for Union Men in Colorado.
Free Speech Denied in Colorado.
Soldiers Defy the Court in Colorado.
Wholesale Arrests Without Warrants in Colorado.
Union Men Exiled From Home and Families in Colorado.
Constitution Right to Bear Arms Questioned in Colorado.
Corporations Corrupt and Control Administration in Colorado.

Right of Fair, Impartial and Speedy Trial Abolished in Colorado.
Citizens' Alliance Resorts to Mob Law and Violence in Colorado.
Militia Hired by Corporations to Break the Strikes in Colorado.

Moyer was released on March 28 and spent that night in a Telluride hotel after officials said he could leave the district. Ten minutes before he was to board a train, a detail of soldiers came and took him to military headquarters. There Captain Wells of the militia notified Moyer that General Bell was detaining him as a military prisoner. He was taken to the Bull-Pen where he cleaned gutters under guard. His attorneys got him a writ of habeas corpus, but the military refused to recognize it. It looked as if it were going to be a long stay.

It did not take long for Moyer to come to terms with his predicament. He was all alone; none of his associates could see him, not even his attorney. He became physically ill and depressed. At this point, he realized that if he did not cooperate with the military officers, he would continue to be treated as an animal. So he began saluting Wells and other officers at each visit, and his treatment became better. During one of their frequent discussions, Bulkeley Wells told Moyer that if he would be a conservative man like John Mitchell, he would be a great labor leader.[42]

After Moyer was arrested and taken to Telluride, Orchard rode horseback to Silverton and somehow managed to reach Moyer by telephone, but accomplished nothing in this way. The murderer remained there for three or four days before he headed north to Federation headquarters. Arriving in Denver, Orchard reported to Haywood the details of Moyer's arrest, which the big man had already learned from telegrams and newspapers. Haywood wrote of the incident, "I didn't like the way in which Moyer had been arrested, with no resistance on the part of his supposed protector." Orchard knew enough to stay away until Haywood was in a more responsive mood, but the union leader never completely forgot Orchard's actions at the time of Moyer's arrest.

On the advice of his friend Pettibone, Orchard took a room at the Belmont Hotel on Stout Street above Pettibone's store. The rooming house had a somewhat dubious reputation because of its clientele, but its location was ideal for him. At the merchant's suggestion, the murderer also used the name of Harry Dempsey whenever he stayed at the Belmont.[43] Meanwhile, matters were becoming much more complicated, and Orchard would soon be called upon to play an important role in the activities of the Federation.

CHAPTER 10.

If Steve Adams had been surrounded with good influences, it is just possible he never would have committed a crime.
. . . James McParland, Pinkerton Dectective

Haywood was arrested several days later. The charge was desecrating the American flag. Upon hearing this, the union leader hastened to the Denver law firm of Richardson and Hawkins to find out what he could legally do to avoid being taken to Telluride by the sheriff. Hawkins suggested that Haywood be charged with the same offense in Denver, and if a friendly judge were to try to case, he could be placed in charge of a law officer while seeking bail and still be allowed to get on with his work.

Haywood hurried to George Pettibone's store to discuss what had to be done. Haywood asked Pettibone to have one of the boys swear out a warrant for his arrest. That very afternoon Jake Wolf swore out a warrant, and it was dutifully served on the union leader. The case went to trial in the courtroom of Judge Billy Hynes, a union partisan. Attorney Hawkins represented the defendant, William D. Haywood, and asked that court to fix bond. Judge Hynes granted the request. Since it would take the defendant time to secure the amount of money stated by the court, the judge released Haywood into the custody of a deputy sheriff. When the specified time limit was up, Hawkins asked for an extension, which Judge Hynes granted without hesitation.

The results of the flag desecration case against Haywood were best explained by the union leader himself:

> When my case came up before Judge Hynes, I came into court with all kinds of specimens of advertising using the American Flag. Unknown friends had sent me these in every mail. I must have had twenty or thirty samples: tobacco sacks, cigar boxes, labels from tomato cans, the flag of a colored men's political club with announcements written on the stripes, and the business card of the Pinkerton Detective Agency, with the all-seeing eye surrounded with flags, the principal among them being the Stars and

Stripes. The case was dismissed. Some time later the use of the flag in advertising was prohibited by law.

After this incident, Haywood and Pettibone were determined to teach the Telluride sheriff a lesson, so Pettibone talked Orchard into teaming up with him to do the job. Orchard was eager to return to the good graces of the Federation, so he quickly accepted, and they planned their strategy. Learning that Sheriff Rutan was leaving Denver on the evening train, they decided to wait near the depot in an alley off Seventeenth Street. When Rutan reached the alley, Pettibone planned to hit him with brass knuckles, after which Orchard would drag him into the alley and kill him. The men waited. The sheriff showed up on schedule, but other men were with him. The unexpected turn of events threw the would-be-killers into a panic, and the intended victim walked unmolested to his waiting train.

Orchard returned to Denver and spent a week, conferring much of the time with Haywood. The union leader was infuriated at the thought of Moyer's being held in Telluride under martial law. The solution was simple: Since there was no justice in the courts, the only solution was to kill Peabody. Since the governor did a great deal of walking back and forth to the capitol from his home, he was a good target for any assassin.

Sleuthing Peabody, Orchard learned his habits and was confident the man could be murdered with ease. His first thought was to shoot the governor, but to do this he needed help. Evidently the failure at Ouray to use his gun to protect Moyer had left its mark. Orchard knew Haywood despised him for that failure. And he despised himself, a professional killer, for being afraid to use his weapon. He did not want to fail again, so he wanted someone with him who would act in an emergency.

Haywood suggested a killer much in the same mold as Harry Orchard, another hatchetman from the Cripple Creek mining named Steve Adams whose specialty was terrorizing non-union miners. Strangely, Orchard, having met him in the saloons of Independence, was already slightly acquainted with Adams. They even lived within a short distance of each other. Orchard was prepared to travel to Cripple Creek to contact Adams, but the only drawback in using Adams, in Haywood's opinion, was that he was too well known in Denver. However, they decided that if Adams moved around mostly at night and kept out of sight during the daytime, he might not be recognized. It was worth a try, so Orchard boarded a train and headed south.

Steve Adams was tall and slim, with light sandy colored hair and grey eyes framed with drooping eyelids. His marked face had the complexion of a man who drank more liquor than was good for him. Adams liked tobacco and boxing, especially when a little blood-letting was involved. Like his companion in crime, Harry Orchard, Adams impressed people as a man who could easily be swayed, but definitely he was not an individual who could organize and plan by and for himself. Of the two men, Adams was considered the more lethal, for he had a more lengthy and notorious record than Orchard at that time. Only a thin line, however, separated their instinctive cold-blooded actions.

In short, Steve Adams was a man of limited abilities, few scruples, and little conscience, and with the capacity to kill at the slightest urging. He had arrived in the Cripple Creek mining district from Telluride in November, 1903, after his ruthless killing of mine manager Arthur Collins. He soon ran into trouble and was arrested for the Vindicator explosion, Orchard's bloody handiwork. He was held in jail and was not released until the Parker-Davis trial on train derailment.

Whatever Harry Orchard conveyed to Steve Adams about the Peabody assassination, it was persuasive and effective. The sleepy-eyed murderer was agreeable, adding "he was ready for anything." When Steve Adams arrived in Denver as planned, Orchard, who had gone ahead, met him at the train depot and took him to the Belmont Hotel. Later they met with Haywood to discuss the job. He gave Adams some money to buy some new clothes and clean himself up. Pettibone also gave each man a sawed-off shotgun, a good weapon since it could be broken down into two main groups and carried around the neck by tying a cord to each part. The gun was hidden by wearing a coat, but when it was needed, it could be quickly assembled.

Pettibone lived a short distance from the governor, and there was a dimly lit street the killers could use. The shot-guns were hidden in Pettibone's yard, next to a vacant lot, and a small trench was dug in the ground for the guns. Pettibone was to remove the guns, clean them, and make sure they were in operating order.

Luck was with James H. Peabody. Time after time the murderers waited for the governor to come home at night, but for one reason or another they never made contact. One night, as they were waiting across the street from the governor's residence, they saw his carriage approach. Steve Adams told the story this way:

> Along between ten or eleven o'clock, or perhaps a little after eleven, there was a carriage drove up to Mr. Peabody's house. We ran out of that yard and ran across the street behind the carriage on the sidewalk. There were three woman got out of the carriage. We had our shotguns in our hands ready for use. Mr. Peabody did not happen to be in the carriage, but the women saw us. They looked at us as the carriage started back, as we saw he was not there, and we started north. Part of the women went into the house, and one remained watching us until we went to the corner of the block and turned east.

Frustrated once more, the murderers considered another means to kill the governor, a more complex method but one that appealed to men experienced in handling explosives. One of the governor's habits was to walk from his home to the statehouse in the mornings between nine and ten o'clock. What could be easier, they reasoned, than to plant a bomb under the sidewalk on Peabody's route? The bomb could be buried in one particular area and a fine wire to set if off could be stretched across some vacant lots to a natural hiding place. The bomb was constructed in Pettibone's store, in the Belmont Hotel Building, and the merchant furnished boards from dry goods boxes to house the explosives.

By now it was the middle of May, 1904, and an event was to take place which forced further postponement of the Peabody assassination. The executive board of the Western Federation of Miners was meeting to make arrangements for the annual convention in Denver. Under the circumstances, Haywood felt that the murder of such a prominent and influential individual as Peabody would cast suspicion on members of the executive board so he vetoed the slaying attempt, and Governor Peabody was given a new lease on life.

Meanwhile, during a meeting of the executive board of the Western Federation of Miners, a crisis occurred when James Murphy of Butte, Montana, refused to sign an emergency financial order which would have provided for the transferring of expenditures out of the union's general fund into the emergency fund. This action would be subject to the discretion of the secretary-treasurer, Haywood.

At the same time, another battle was being fought. It had its origin several years before. In July, 1903, a strike that had been waging for several weeks in Idaho Springs, Colorado, reached a turning point when the big mines in the district began to operate with non-union men. The last mine to go into operation was the Sun and Moon mine. Late at night on July 28, the night watchman saw two men roll kegs of explosives down the hill into the mine's transformer house. An object from the resulting explosion was thrown with such tremendous force that it pierced the body of one of the Federation's dynamiters and killed him.

The incident so aroused the citizens of Idaho Springs that they formed a Citizens Alliance Committee, rounded up all the Federation members they considered dangerous, and ran them out of town. Soon after this, fourteen Federation members were tried for the destruction of the Sun and Moon mine transformer house. A private detective by the name of Lyte Gregory was hired by the prosecution to gather evidence in the case. He was also one of those responsible for deporting the union miners. All fourteen defendants were acquitted, but Gregory's involvement earned him the hatred of the Federation. As a result he was a marked man, and it was to cost him his life.

Foster Milburn, a Federation member from Idaho Springs, happened to see Gregory the very morning that the detective arrived in Denver after visiting his brother in La Salle, Colorado. Milburn immediately reported Gregory's arrival to the Federation and then consulted with George Pettibone. Pettibone was unusually close to the Federation miners from Idaho Springs, many of whom were staying at the Belmont Hotel and were using his store as their headquarters.

After meeting with Milburn, Pettibone contacted Adams and suggested they should either kill Gregory or mutilate his body in some manner. Mutilation was chosen since it was claimed that Gregory had done the same thing to an old man during the height of the Idaho Springs troubles, an unfounded accusation. Pettibone's obsession to see Gregory dead pushed him to get something moving. He tried to stir Adams into action, but he seemed reluctant. A self-starter like Harry Orchard was needed, they both decided.

Seeing another opportunity to atone for his failure in Ouray, Orchard immediately pledged his support. Returning to the store on Stout Street, the three conspirators made plans for the detective's death. After that, they went to Federation headquarters, and while Pettibone went inside to confer with members of the executive board, Orchard and Adams waited. Even though some individual board members expressed an opinion that Gregory should receive his just dues, there seems to have been no direct orders for his execution, but there was also no attempt to call off the mad dogs. It was purely a Pettibone-instigated crime. He pulled the strings on his puppets, Orchard and Adams, and they killed for him.

After he arrived in Denver, Gregory visited the office of his employer, the Reno Detective Agency in the bank building. He then went to Mart Watrous' saloon on Curtis street for a drink of bar whiskey. Leaving the Watrous saloon, Gregory went on to visit three more bars all located on the same block on Curtis Street and spent the entire afternoon drinking.

The real action began in the early evening when Pettibone, Orchard, and Adams found their intended victim. Entering the bar, they saw the detective and an old friend, "Doc" Floyd, and Milburn drinking in the back. There, they studied the victim as Milburn identified him. Satisfied, the three men left the saloon, walked directly across the street to a more advantageous spot, and waited for Gregory. After a short time Pettibone made an excuse to leave, saying he would return in a few minutes, but he never did.

At eight o'clock Gregory and Floyd left Walker's saloon. Floyd urged his friend to go home, but the detective retorted that he was not in the mood. At Gregory's urging, they boarded a street car and rode to Third Avenue and South Eleventh Street where they got off and walked to Ed Clary's saloon. Seeing the two men leave Walker's saloon, Orchard and Adams quickly boarded the same street car as their victim. Milburn did not join them at this point as he sought out friends in an effort to establish an alibi for himself.

In Clary's place, Gregory and Floyd sat down at a game table in one corner and began playing pitch-for-drinks with two other men, later switching to poker. Milburn had rejoined his two confederates in the "stake-out" outside the building. Interestingly, none of the three assassins carried any type of weapon, making it necessary for Orchard to return to his room for a gun. He selected a sawed-off pump shotgun, which he broke down and hung around his neck, and thus concealed the weapon under his coat.[44]

The original plan had been to shoot the detective through the window of the saloon, reminiscent of Adams' slaying of Collins. But the card game unexpectedly moved to a room in the rear of the structure, and the victim was now out of view and effective range. Temporarily stymied, Orchard brazenly entered the saloon, went to the bar, and ordered beer. He saw the door to the game room was closed. This left the alternative of waiting until Gregory came out of the saloon so that they could hit him with buckshot out in the open.

The murderer walked out of the saloon and across the street into the shadow of a large tree where he explained the situation to his accomplices. A little after midnight, the unsuspecting detective, feeling the effects of the liquor, arose from the card table and announced that he was going home. Staggering to the door, he inquired about directions. A man by the name of Baldwin walked over to the drunken man and told him to walk straight up Third Avenue to Sherman Avenue, and he would soon be home. Gregory started out the wrong way but Baldwin came to his rescue and turned him around.

The three men watched the drunken man leave the bar; they crossed the street to be on the same side. In their haste, they stumbled over some wires stretched on the outside of the sidewalk to protect the lawns. Startled, Gregory stopped.

Orchard fired, but the bullet missed. The victim turned around and shouted, "You damned murderers!" In the same instant, he reached for his revolver, a .45 caliber Colt, which he carried in a holster hung from his shoulder on the left side. He got the weapon out, but before he could fire it, Orchard shot again. The buckshot hit the detective's right hand and hip, knocking the Colt from his grasp. The murderer fired once more, and this time the shot severed an artery in the left side of Gregory's neck and killed him. The big man collapsed and lay motionless.

On Sunday afternoon, the day following the slaying, Orchard and Adams went to Jack Simpkins' room. The executive board was not in session, although a meeting had been held in the morning. Simpkins was waiting for them, anxious for details. Orchard felt close to the man from the Coeur d'Alenes. He knew Jack had served time in the Bull-Pen after the 1899 riots and had been harshly treated.

The morning papers were out, and an account of Gregory's slaying appeared. Haywood and Pettibone joined the trio to rehash the murder. Haywood seemed particularly pleased with the job, but then he always seemed pleased with a succesful use of violence. The big one-eyed man told them that he had contacted a police official and had been told that whoever "bumped-off" the Idaho Springs detective had done a pretty good job. It had been intimated that the police would not look very hard for the killers. Milburn had already been questioned by the police, but had not been arrested. His alibi had apparently held, and the crime was to remain officially unsolved. Adams and Orchard were given a hundred dollars each by Haywood for the murder, along with a little drinking and gambling money. More important for Orchard was that he felt he had redeemed himself.

According to Haywood, the killing had provided a solution to the crisis that was tearing the executive board apart. That very morning a few members of the executive board had met in the Federation offices. Simpkins had taken a copy of a local newspaper containing an account of the slaying and casually handed it to James Murphy, the board member who had refused to sign the emergency order. Murphy's face drained of color; his lips were tightly compressed as he read the article. Haywood then handed him the emergency order and told Murphy he had the honor of being the first member to sign his name to it. When the man from Butte hesitated, Haywood's voice hardened. He told Murphy he would not only sign the order, but he would also damn well like it. The issue was resolved, and Murphy signed. A large amount of money could now be taken out of the treasury and placed in the emergency fund, for Haywood's use. Haywood also indicated that if Murphy had not signed the order, he would not have left the room alive.

At the time that Lyte Gregory was killed, Orchard and Adams were living in a room over a saloon at 38th Street and Downing Avenue. The saloon was a convenient drinking establishment and kept Adams from having to wander too far from his room. Frank Gylester, a former policeman, ran a saloon across the street and knew the two killers as regular customers, but Orchard was using the name of Hogan and Adams was known as Dixon.

Both men were heavy drinkers but Adams, or Dixon as he was called, was the heavier. Orchard did not like to drink while he was on a job, but Adams was not that prudent. Both gambled, but Orchard seemed to prefer the cardboards. Both chased prostitutes, but in the presence of the women Orchard was the more boastful of his exploits, real and imaginary.

Orchard and Adams later moved to a rooming house on West Fourteenth and Santa Fe. One night when they were drinking in a saloon run by a Frenchman, Orchard became boisterous and offensive. Some of the customers, amused by his antics, tried to pick a quarrel. Orchard's quick temper surfaced, and he pulled a gun. As he left the saloon, he threatened to shoot the first one who made a move.

A short time after the Gregory killing, Adams went on a monumental drunk, and his friends, fearing he would get into more trouble, arranged to send him back to his wife in Cripple Creek. Orchard, however, preferred to stick around Denver for awhile and see if he could unearth some more excitement, and he soon discovered that it was not hard to come by as the Federation pursued its violent course.

CHAPTER 11.

Study the life of this man; the commission of crime with him was supreme over everything else in life . . .
. . . William E. Borah

Trouble was brewing. On May 23, 1904, when the annual convention of the Western Federation of Miners met in the Odd Fellows Hall in Denver, it began as a loose affair, argumentative in nature. Many of the delegates were deeply concerned over the large amount of money being spent on strikes, and there was talk of electing new officers; the delegates were bickering and fighting among themselves. The atmosphere was tense and strained. The Federation seemed to be coming apart at the seams from within.

Even the leadership of the union was showing a difference of opinion as to their real objectives.The widening breach between Moyer and Haywood was becoming evident. There existed, first of all, a great deal of jealousy between the two officials. Bill Haywood was a natural leader, a forceful pusher when he wanted something, and the fact that he was just secretary-treasurer and not president was usually overlooked. Moyer, on the other hand, was an able administrator capable of providing sound leadership. He lacked, however, the force and dynamism of Haywood. Haywood had a flair with groups, and once he was described as a man who could make himself understood by a crowd that didn't understand merely by waving his arms and shouting.

When Moyer was in jail in Telluride, he sent his business correspondence to his wife, who in turn would get a member of the executive committee to handle it. Needless to say, this snub rankled Haywood. When talk filtered back from Telluride that Moyer was not only coöperating with the military, but also showing them respect in the process, Haywood was infuriated. Meanwhile, Moyer was beginning to have enough of being arrested and confined. There must be a better way. To Haywood, there was no other way. He even talked periodically of eliminating Moyer. It was becoming a case of two determined men fighting to see who was going to take charge of the sprawling union.

At the convention, Moyer's absence was noticeable. He was the unseen representative of every Federation miner who had ever been in jail or had spent time in the Bull-Pen because of union activities. Prior to the opening of the convention, Moyer had arranged with Pettibone to have his picture hung on the platform over the president's chair. For added effect, Moyer requested that when the delegates were photographed, a vacant chair was to be put in the president's place. Thus, even while he was incarcerated, Moyer fought to maintain his official position.

From the very first day, the convention threatened to break up into splinter groups. Every night different factions held meetings, and there was even open talk of replacing Haywood. Something spectacular was needed to unite the Federation into a working unit, and that action would have to be of such importance that it could not be ignored.

Pettibone, Davis, Parker, and Orchard spent a great deal of time considering the problem. They agreed it would take a powder job of dimensions great enough to whip the state of Colorado and the mine owners in every western state into submission. Things had been too quiet in the mining districts. The obvious solution was to stir things up and get the union moving by scaring the opposition with shattering viciousness and violence. They must get public sympathy by being unjustly persecuted.

Above all, a show of power had convinced Haywood that if he wanted support, financially and numerically, such violent action was necessary. He once mentioned to Orchard that the riots in the Coeur d'Alenes had cost the union those local organizations, but the gains made in other western states justified the action. This seemed to convince Haywood that the best way to sway the masses was the ruthless use of power. This decision was to have a most profound influence over the direction he took in union leadership.

The something "spectacular" the union men talked about would have to stir up the delegates, unite them, end their quarreling, and finish their business. All of this was essential for the Haywood faction to remain in leadership roles. A course was then charted on those premises, one that would most certainly be an act of desperation, showing that the most vicious and disruptive acts inevitably spring from necessity.

It was generally agreed that the time for the big spectacular had come. A previously suggested target was the destruction of the Independence depot of the Florence and Cripple Creek Railroad. The station was closed, but the platform was still used by a number of non-union miners who commuted to the Findley mine. The idea had been previously discarded because of the magnitude and impact of such a deed, but now the destruction of the station was reexamined in more positive terms. Orchard argued that the structure should be blown when the non-union miners were making a shift change. The night shift would be ideal. The longer the group debated about the possibilities and the results, the more the plan was favored.

Haywood was definitely in favor of the plan. He gave Orchard money, but he also expressed concern about Orchard, urging that he not get mixed up in the actual destruction of the station. He should let someone else do it. The killer stated that he would do just that, or he would detonate the charge by using an alarm clock. At any rate, Orchard promised he would not be in the vicinity of the explosion when it occurred.

The prospects for committing murder on such a large scale were terrifying. They were going to kill a great many men. At times, the prospect almost overwhelmed him. Orchard, on his return to Cripple Creek, found the district to be unusually quiet. It had been that way since the Vindicator explosion. It seemed so orderly that the governor was considering a proposal of withdrawing state troops.

Orchard also found equally astonishing a unique innovation — the Federation was doing a great business. They had started union stores, at first out of necessity. The strike in Cripple Creek had been only three days old, when on August 13, 1903, the business men of the mining district, pressured by members of a citizens' alliance, an anti-Federation group, announced that from that day on, business would be conducted on a cash only basis. The union responded by bringing carloads of potatoes, flour, and other commodities into the district for the striking miners. To distribute this strike relief, Haywood suggested that they establish stores of their own. Actually, it was a spin-off on the old company store, an institution that had plagued mining camps for generations.

Nonetheless, the union stores were a quick success. Another feature that was used to good advantage was the coupon book. Each paper book contained coupons of different values, good at face value for anything in the store. When Orchard returned to Cripple Creek toward the end of May, there were four such stores in operation.

Orchard had a lot on his mind when he came back this time. He was starting a new job, and he was looking for a partner. He needed not only physical help in setting up the death trap, but also moral support. Haywood suggested using Steve Adams, but Orchard hesitated because the murderer drank too much and didn't know when to stop, no matter how important the job. So he passed Steve Adams in favor of Billy Aikman. Aikman, however, begged off since he had purchased an interest in a saloon and tended bar at night. He reasoned that he might be missed if he were not on duty, and a man with his reputation could hardly afford to be missed.

Not many men had the intestinal fortitude to attempt to kill a large number of human beings in one fatal swoop. So Orchard was limited in his choice of human material, and had to return to Adams. Would Steve Adams help him with a small job? Yes, he would. And so they got together and started planning.

First they had to get explosives. Floyd Miller was working a small mining lease and could easily get the killers a hundred pounds of dynamite and two boxes of blasting caps. Miller would also know better than to ask what the powder was going to be used for. Orchard paid him for the explosives and arranged to pick them up several days later.

The next consideration was an alibi. Orchard finally decided to use a camping and hunting trip to Wyoming with John Neville as the perfect screen. The pair had talked about such an excursion often, and Ida Orchard was under the impression her husband was sincerely taking the trip to find work and settle down. The men could make a production about leaving town, and then Orchard could double back at night and blow up the station.

Neville had received several hundred dollars in insurance money when his saloon had burned, and he gave his friend Harry a hundred.[45] Money was no problem in financing the trip. Neville was also unemployed, so there was no reason not to go. Besides, he was still in debt to Orchard for burning his saloon. Adding a special touch, Neville was innocently persuaded to take his teenage son, Charley, along. Now, it would definitely appear to be just a couple of men and a kid off on a camping trip, and only Orchard knew what was really going to happen.

On Thursday, June 2, 1904, Steve Adams and his wife called at the Orchard's home. After the usual small talk, Adams got up and took off his coat and vest, saying to Mrs. Orchard, "I wish you would let me have your brother's mackinaw." He was referring to Joe Crieg's coat, which he had left when he visited his sister. Ida Orchard supposed Joe would not mind, but she could not imagine why Adams wanted the coat. Obviously, he wanted to avoid recognition. Ida Orchard asked the two men where they were going, and in a joking manner one of them replied, "We're going to see some girls and will be back in an hour or so." Laughing, the killers left the house. The women assumed they were stepping out for a drink.

At Floyd Miller's house they removed the two fifty-pound boxes of dynamite and the two boxes of fuses from the rear of the cabin and carted them to a cabin on the hillside above Steve Adams' former house, which now belonged to his brother, Joe, and was empty at the time. The explosives were placed on the floor and the door securely locked. Adams put the key to the lock in his pocket, and they returned to Orchard's house, arriving about midnight, where they changed clothes. Adams and his wife left the Orchard home, and the women were never aware of their activities.

Orchard observed the habits of the non-union miners as they went on shift and then returned. He carefully checked all schedules and routes. This was to be his greatest feat as a professional assassin. He was aware of its importance to his status and his career. If this job were successful, he would not have to work for the rest of his life. It was a job to Harry Orchard, no more, no less. The human element involved did not concern him.[46]

Neither Neville nor Orchard had an adequate means of transportation for the intended trip, so they bought a team of horses and a wagon from Joe Adams, Steve's brother, and made final arrangements to leave on Saturday afternoon, June 4th. According to plans, Orchard was to return that night and help Adams destroy the station.

On Friday, the day before the hunting trip, Bill Easterly visited Orchard at his home and said Parker had just arrived from Denver and wanted to see him; it was important. The killer went to Parker, who also lived at Independence, only to learn that a new wrinkle to the plan had been thrown in. The convention in Denver had appointed a committee to investigate the Cripple Creek strike. One frightening objective of the committee was to meet representatives of the mine owners and hear their side of the story.

Haywood had been against the appointment of the committee in the first place since it might prove unfavorable to him if the committee report conflicted with his own position. It might prove more embarrassing if the committee were somehow to effect a strike settlement. In short, he wanted the project to go on as planned, but not while the committee was in the district. Orchard agreed to wait until the committee left on Sunday.

An unusual conversation took place on Sunday while Orchard was scouting the area of the Independence depot to check it out in case any last minute adjustments might need to be made. While there, he ran into a friend whom he had known for several years, and the conversation quite naturally drifted to the district's labor problems. Orchard, for some unexplained reason, commented that, "Something terrible is going to happen tomorrow." Interested, his friend pressed him for further details, but the killer would only look at the ground and shake his head. "I can't tell you what it is," he said, "but mark what I'm saying: something terrible is going to happen tomorrow." Perhaps Orchard had some inner doubts about the project, or perhaps he just wanted to be a prophet of doom. At any rate, he was not above boasting and constantly trying to overawe people with his importance.

According to Mrs. Neville, her husband and Harry Orchard left the town of Independence about 11:30 a.m. on Sunday, June 5th, headed for Colorado Springs before going north to Wyoming. After a few miles of riding, they stopped to make preparations to camp for the night. As darkness settled, Orchard told Neville he was going back to see a party about a matter he'd forgotten about. If anything happened while he was gone, Neville was to swear that he had spent the entire night at the campsite.

Shortly after dark, Orchard saddled one of the horses and headed back to Cripple Creek. The killer stopped his horse just a mile short of the Independence depot and tied the reins securely to some brush. Then he walked the rest of the way to the cabin where the dynamite was stored. Adams was waiting. It was almost ten. The two men immediately began constructing a small windlass out of wood, to be used to detonate the dynamite charge. With a strip of leather, two small vials were then fastened to the cross-piece of the windlass. A wire was then attached to the windlass. When the wire was pulled, it would upset the windlass and the vials would spill their contents of sulphuric acid on the blasting caps imbedded in the dynamite. After that, the station would be blown sky high.

A few minutes past eleven, each man picked up a box of blasting caps and put them in his coat pocket. Then each man picked up a fifty pound box of dynamite and put it on his right shoulder in a carrying position, and they were started for the deserted station. As they approached the empty depot, the night made it appear much larger than usual, and its light color gave it an awesome look. For some strange reason, Orchard had the feeling that he was entering a grave yard. The depot itself had been closed for some time; it was just a place where the train stopped. The station was built on a side hill with a long platform constructed in front of it, and to the advantage of the killers, the area was poorly lighted. It was, as one man commented, "a spot peculiarly suited to the perpetration of crime and deeds of darkness."

The killers, carrying their cargo of death, struggled under the station platform. Orchard selected a central spot where the wooden boards of the platform were close to the ground. He put his load down, and wiped the grimy sweat from his face with his sleeve. Scraping away the earth, he dug a hole large enough for the two boxes of powder, then put them into it. The top of one box was left exposed, and the blasting caps were placed in the powder.

Adams later stated, "We sprinkled powder composed of burnt sugar and chloride of potash in those caps which were open end up. We set the cap box into the box of powder, after preparing a hole in the powder big enough to set it in, then put a windlass on the box with two bottles of sulphuric acid." This was the tedious part of the task. Using a piece of tin can, Orchard placed it under the windlass, covering the dynamite caps. This precaution would offer protection from an accidental explosion if any of the acid happened to spill. Taking a deep breath, the killer then filled the small vials on the windlass with sulphuric acid from another bottle.

There was no light, and he was cramped in the small working area. A wrong move, the slightest spillage of the acid, and it could be over. Beads of sweat formed on his nervous face. His hands were damp, and he could taste the salt water from his face as it rolled over his lips. But his hands were steady, betraying none of the inner tension. Harry Orchard never forgot that moment — it was "ticklish business," he later confessed, as the acid flowed into the vials without a mishap.

Wire was then stretched from the windlass to a spur track two hundred feet distant, and the free end was fastened to a chair-rung by Adams. That chair-rung was Steve Adams' contribution to the plan. Unfortunately, a fellow by the name of Arthur Rogers had seen Victor Poole get the chair-rung from Zingler's saloon and give it to Adams and Parker. Rogers remembered the incident because he thought they would use it to beat someone up.

Satisfied that everything was ready, Adams and Orchard entered an old ore-house near the spur track to rest and wait. They were calm and collected as they silently waited. It had been a dark night, but about two o'clock the heavens began to lighten and the stars came out. Voices were suddenly heard from the station as the miners gathered to await the arrival of the train. Occasional laughter came to them, but mostly what the two men heard was the tired small talk of men working the night shift.

About 2:15 the engine lights of the Florence and Cripple Creek Railroad were seen. In the ore-house, the murderers heard the sound of the approaching train. The engine was within one hundred feet of the depot, whistling its call of arrival. At this moment, there was no hesitation for the two men. There was a job to be done, and they were professionls. "After the train stopped, Orchard and I each took hold of the rung of the chair, which passed through the loop of the wire at our end, and both of us pulled it and caused the explosion," Adams stated.

The acid spilled on the caps and dynamite, causing an instantaneous fire and an explosion. A loud, thundering, terrifying noise signalled that the station platform was disintegrating beneath the unsuspecting men. Suddenly, the sky filled with rocks, dirt, timber, and bodies. The explosion was raining death over the surrounding countryside. Where the platform had been was a gaping hole in the torn ground. The area was a mass of broken and shattered lumber. The station was wrecked — floors were blown up, the roof was pierced, and the walls were splintered. Like desert sagebrush, heavy pieces of timber littered the ground. When the explosion occurred, the platform was lifted and torn apart, while the building itself heaved, shook, and reverberated with the concussion, and flying projectiles riddled the structure.

When the rain of debris stopped, the cries of the wounded emerged. The sound and force of the explosion awakened people in the vicinity; some were reported to have been thrown from their beds by the force. Ida Orchard woke suddenly, absolutely terrified. She thought at first that lightning had struck the house as a flash had come in over the door. Her house was close enough to the death scene for her to hear the groans and screams of the survivors. Her next thought was that there had been a mine accident.

Soon the ghastly, bloody spectacle was crawling with people. The bodies of the men, many blown a hundred and fifty feet straight up the hillside, were beyond recognition. Parts of human bodies lay strewn grotesquely over the area — a hand there, a foot here. It was all so unbelievable.

Thirteen men were killed that fatal night of June 6th. At least six amputations were necessary. Pieces of flesh and bone were gathered up in tin pails, baskets, buckets, and barrels. In some instances the dead could be identified only by their hands or their clothing. Phil Chandler related that the woman with whom he boarded "went down to the undertakers and picked up one man's hand in a barrel full of remnants in the back room there." The day following the mass slaying, Steve Adams took his wife and Mary Ann Mahoney, with whom they were living, to see the scene of the tragedy. "We looked down at the depot, and it was a hard sight to look at. I did not think it would be so bad," he was later to remark.

When they were planning the crime, the perpetrators had fully intended to take the chair-rung and attached wire with them after the explosion. But when the powder charge went off, the debris falling around them was so heavy and intense that they broke and ran away as rapidly as they could move. Orchard claimed they followed the railroad tracks nearly to the old Victor mine on the north side of Bull Hill, then separated. Orchard had conveniently left his horse near the Colorado Springs road. Since the route to the camp site was down hill, he could ride his horse full speed. It was a chilly night, and since the horse was cold from standing so long, he made record time. Both the Nevilles were awake when Orchard arrived in camp shortly after three. John got him aside and asked him what he had blown up. He'd heard the report of the explosion and the shaking of the ground, and Neville had spent too many years as a miner working with powder not to know what had happened. Orchard was evasive, but Neville pressed to know it if were the Findley mine. Orchard reiterated that he was not near the explosion, so he didn't know what had happened. Neville dropped the subject, and Orchard crawled up into the wagon and got into bed. He had done a good night's work and had successfully completed a job that would cause repercussions far, far beyond the site of the actual blast.

Fig. 2. Silver City, Idaho, miners. Haywood 4th from right on front row.

Fig. 3. Bunker Hill and Sullivan power plant, mills, and office at Kellogg, Idaho.

Fig. 4. Ruins of the Bunker Hill and Sullivan concentrator after the explosion.

Fig. 5. The "bull pen" of the Coeur d'Alenes — aftermath of the riots.

Fig. 6. Independence (Teller County), Colorado.

Fig. 7. Arrival of soldiers and deputies with prisoners after the Battle of Dunville, 1904.

Fig. 8. One of Harry Orchard's bombs: lead bucket, windlass, and vials for chemicals.

Fig. 9. Scene of the explosion on the 600 foot level of the Vindicator Mine where on November 21, 1903, Charles H. McCormick, superintendent, and Melvin Beck, shift boss, were killed.

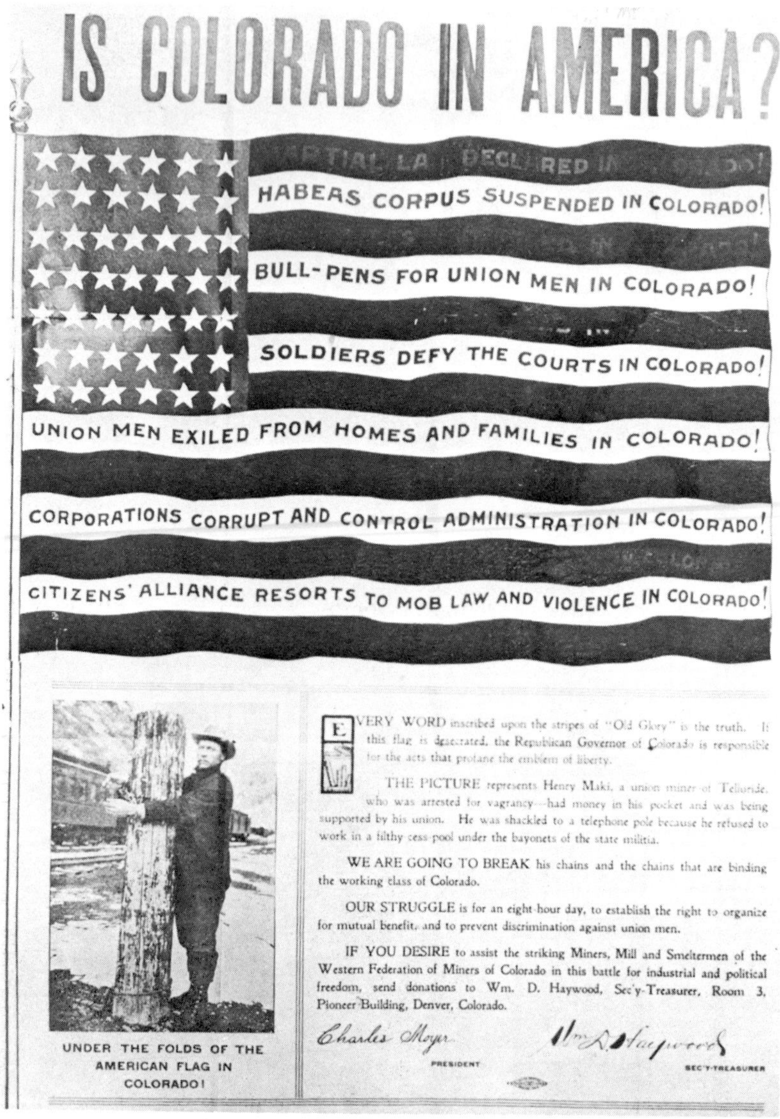

IS COLORADO IN AMERICA?

HABEAS CORPUS SUSPENDED IN COLORADO!

BULL-PENS FOR UNION MEN IN COLORADO!

SOLDIERS DEFY THE COURTS IN COLORADO!

UNION MEN EXILED FROM HOMES AND FAMILIES IN COLORADO!

CORPORATIONS CORRUPT AND CONTROL ADMINISTRATION IN COLORADO!

CITIZENS' ALLIANCE RESORTS TO MOB LAW AND VIOLENCE IN COLORADO!

UNDER THE FOLDS OF THE AMERICAN FLAG IN COLORADO!

EVERY WORD inscribed upon the stripes of "Old Glory" is the truth. If this flag is desecrated, the Republican Governor of Colorado is responsible for the acts that profane the emblem of liberty.

THE PICTURE represents Henry Maki, a union miner of Telluride, who was arrested for vagrancy—had money in his pocket and was being supported by his union. He was shackled to a telephone pole because he refused to work in a filthy cess-pool under the bayonets of the state militia.

WE ARE GOING TO BREAK his chains and the chains that are binding the working class of Colorado.

OUR STRUGGLE is for an eight-hour day, to establish the right to organize for mutual benefit, and to prevent discrimination against union men.

IF YOU DESIRE to assist the striking Miners, Mill and Smeltermen of the Western Federation of Miners of Colorado in this battle for industrial and political freedom, send donations to Wm. D. Haywood, Sec'y-Treasurer, Room 3, Pioneer Building, Denver, Colorado.

Charles Moyer
PRESIDENT

Wm. D. Haywood
SEC'Y-TREASURER

Fig. 10. The "desecrated" American flag.

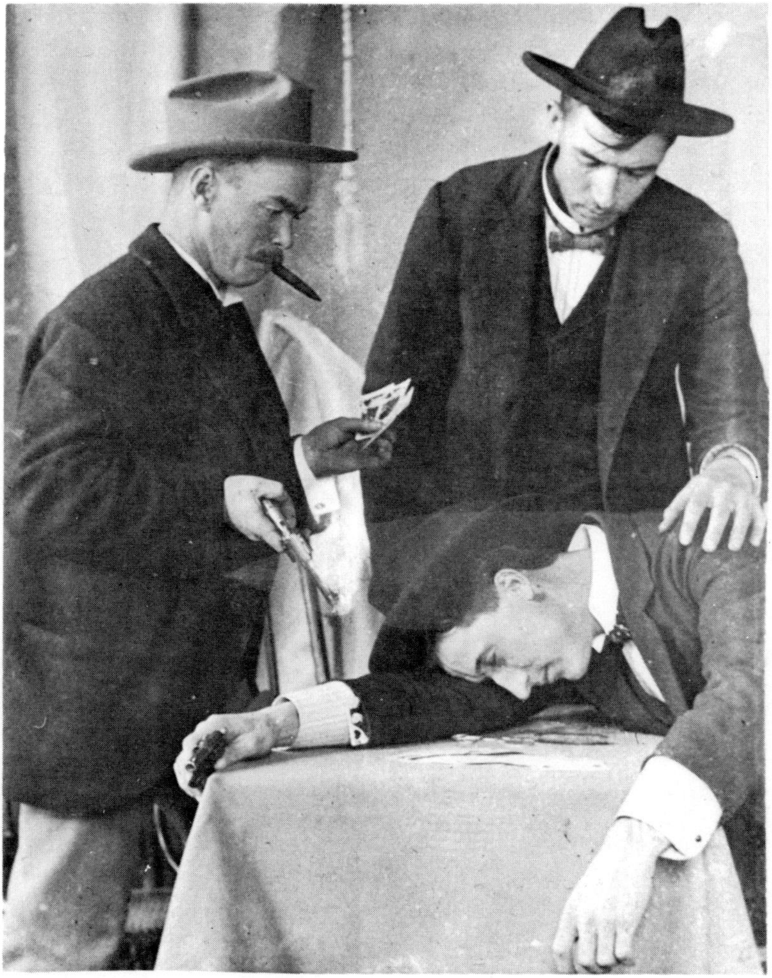

Fig. 11. Harry Orchard shamming a killing with friends in Salt Lake City.

Fig. 12. Victor, Colorado, scene of the first deportation of union miners and other union men by the military, after the June 6, 1904, riot. Second floor of the building on the right was the "bull pen." Deportees are entering the train cars on the left. The wives and relatives in the crowds looking on.

Fig. 13. Front view of the Independence depot after the bomb blast.

Fig. 14. Side view of the Independence depot following the accident.

Fig. 15. John Neville.

CHAPTER 12.

We done a good job.
. . . Jack Simpkins

Orchard desperately hoped Neville would not ask too many questions about the explosion. The following afternoon, they rode into picturesque Colorado Springs and past the stately houses of the men who had made fortunes from Cripple Creek gold, rising from miner to millionaire. Their first act in the city was to purchase a newspaper and hurriedly read, only to find out that the first reports on the disaster had been grossly exaggerated and that none of the local residents living near the depot were seriously injured. Relieved, Neville telephoned his wife and learned that a piece of plank had been tossed through the air by the force of the blast and had smashed through the roof of the Neville house; Mrs. Neville had been struck on the breast as she lay in bed. She was not severely wounded, but she was pretty shaken by the experience. No doubt Orchard held his breath during the conversation. Too much was riding on what Mrs. Neville told her husband. Satisfied that conditions were not extreme enough to justify returning, Neville agreed to continue the trip and asked no more questions, although he could probably guess with little effort what might have really happened.

They finished buying supplies and continued a few miles outside the city to camp that night. Early the next day, Orchard returned to town to buy a morning edition of the newspaper, hoping to get some up-to-date details of the happenings in the distressed mining district. Learning nothing new, he hurried back to camp. With the Nevilles, he started northward and camped a few miles beyond Palmer Lake. By the following afternoon they reached Englewood, a suburb of Denver. Leaving the horses and wagon at a livery stable, the trio took a street car the six or seven miles to the city center.

Here Orchard left the Nevilles and went to see Jack Simpkins in his room at the Granite Hotel at 15th and Larimer Street. He met Simpkins on the street near the hotel, and after exchanging the customary greetings and handshake, the men went up to

Simpkins' room for a more confidential talk. Soon Haywood and Pettibone dropped in, and still later Steve Adams joined them. Simpkins succinctly summed up the feelings of the group when he told the mass murderers, "We done a good job."

Orchard was a little surprised to see Adams in Denver, but he understood after he heard Adams' story. It appears that after Adams had taken his wife and Mary Mahoney to see the wrecked depot, he ran into friends who informed him that he was suspected of blowing up the station and that there was talk of hanging him. The mass murderer quickly returned to his home in Midway long enough to pick up his Winchester, a revolver, and a belt of cartridges. Heading north out of the Cripple Creek district on foot, Adams spent the night at Divide with a family named Pobst, and when he started out the next morning he left behind his rifle and cartridge belt. At South Platte he jumped a freight train bound for Denver. As it turned out, he had left the mining district none too soon; a large mob had gone to his house with the intention of lynching him.

In the union offices Adams was greeted soundly by those people he knew and was introduced to those with whom he was not acquainted. It was soon quite apparent that not everyone knew who was responsible for the recent disaster, showing clearly that not everybody in a leadership or official position was aware or at least in agreement with what a few of those in supposedly responsible positions were up to in their direction of Federation activities. Obviously, a small number of persons in strategic positions can almost control the activities and the destiny of a large organization.

In Simpkins' hotel room Haywood asked the two killers how much money they wanted for the job, reminding them that it would be better if they didn't take all of the money at one time. Orchard said he wanted three hundred, an amount which Pettibone delivered the next day. Adams wanted two hundred, which he also promptly received. Adams said he had no plans for the present, but he thought he would send for his wife and stay around Denver for awhile until things quieted down. The next few days, Orchard and the Nevilles' shopped around for supplies for their projected trip into Wyoming.

On Sunday following the Independence depot explosion,

Orchard met with Haywood in George Pettibone's back yard. The big, one-eyed labor leader estimated that the mass slaughter at the station was worth thirty to sixty thousand dollars to the Federation. Contributions to the union strike fund always went up after some violent incident. Haywood felt good about the whole thing and was convinced that the explosion was the only thing that kept the Federation together. Haywood wrote of the effect of the tragedy on the convention:

> Many of the delegates gathered in the hall early on the morning of June the sixth. Some had newspapers grasped in their hands; others had the papers spread out before their eyes. A horror-stricken look was upon the faces of them all; they were reading about the explosion that had blown up the Independence depot at Cripple Creek the night before.

As soon as the roll call was complete, the Cripple Creek incident was taken up as a special order of business. Few knew anything about the explosion except what they had read in the newspapers. The convention did decide to offer a reward of five thousand dollars for the arrest and conviction of the parties responsible. It was money saved, as no one was ever brought to trial for the mass slaying. Organized resistance toward the labor leaders was now forgotten, and Moyer and Haywood remained in their elected positions.

News of the explosion at the Independence depot spread rapidly throughout the Cripple Creek mining district, causing great excitement and anxiety. All hell was about to break loose. Word was sent to James Murphy, manager of the Findley mine, to the principal stockholder of that mine, and also to a regional political power, A.E. Carlton, and Sheriff Robertson. At three o'clock, a special train left the town of Cripple Creek with those three men and a team of deputy sheriffs, doctors, and nurses. Upon their arrival at the death scene, they found the injured men being well-treated; the bodies of the dead were being removed to the coroner's office at Victor. State militia companies in the district were being notified and ordered to assemble at the Victor armory to await orders to go on state duty again.

The blast sight itself was roped off, and a search for evidence was conducted. The central location of the explosion was found with little trouble, since the hole where the explosives had been stored and detonated was unmistakable. A steel wire extended from the vicinity of the crater, under the broken and ripped station platform, to a distance of approximately two hundred feet. Near the old Delmonico ore-house they found the end of the steel wire still attached to a chair-rung, revealing the method used to detonate the deadly explosion.

Most of the mines closed down, and few men reported for work. By now, it was the general consensus in the district that the men of the Western Federation Miners were responsible in some way for the slaughter at the station. The situation became extreme, with emotion giving way to action. The confrontation with members of the Federation occurred in the hotbed of unionism, the town of Victor.

The county commissioners held a special meeting at that location and decided to call in a stronger and more competent law officer than Sheriff Robertson. Also holding meetings and reaching the same conclusion, were the Mine Owners Association and the Citizens' Alliance. The course of action was to take Sheriff Robertson to the militia armory and demand his written resignation. When he refused, a noosed rope was shown him, and he was told in no uncertain terms that if he refused to sign, he'd be thrown, along with a rope, to the crowd that was gathering outside the building. Robertson needed no further urging, and the county commissioners immediately appointed Edward Bell as sheriff. A number of deputies, the choice of the mine owners, were also appointed.

In the early afternoon on Monday, the sixth, the mine owners held a meeting at a vacant lot in the center of the Victor business district. Across the street on one side of this lot was the union store; on the other side was the miners' union hall. An ugly, ill-tempered crowd gathered. An empty wagon was moved on to the lot to be used as a speakers' platform, and Clarence Hamblin, secretary of the Mine Owners Association, climbed on the wagon to address the crowd. He minced no words and hid no feelings when he condemned the Western Federation of Miners for the depot tragedy and called them a bunch of cutthroats. He went so far as to urge the citizens to take the law into their own hands. A question was shouted by a union miner, and without warning the man next to him swung around and struck him in the face. A fight started, followed by shooting, confusion, bedlam, cursing, and the spilling of blood.[47] Some of the gunfire seemed to come from the union hall, as militiamen cleared the vacant lot in a sweep operation. Five men lay on the ground, two of them fatally wounded.

The union hall was surrounded as Sheriff Bell, accompanied by the postmaster of Cripple Creek, entered the hall and demanded that its occupants surrender. The union men refused, and the soldiers opened fire. Volley after volley poured into the building. The fighting finally subsided, and the union men surrendered; four were wounded. A mob rushed into the hall and completely wrecked the interior. By night time, every union hall and cooperative store in the mining district had been entered and wrecked. Armed groups of citizens, deputies, and soldiers wandered through the district and arrested Federation men. By morning, nearly two hundred men had been imprisoned.

In the Victor Union Hall, a large number of photographs of various mine groups were found. On the back of the Vindicator group, two names had been scratched out — Charles McCormick and Melvin Beck. The military and the mine owners considered this direct evidence of a systematic conspiracy and assassination plot. Federation officials, however, insisted the photographs were used for "scab" posters that were published and circulated at frequent intervals.

On Tuesday, June seventh, Lieutenant Governor Haggott, in the absence of Governor Peabody from the state, issued the dreaded yet inevitable proclamation:

> I, Warren A. Haggott, acting Governor and Commander-in-Chief of the military forces of the State of Colorado, by virtue and authority in me vested, do hereby proclaim and declare the said county of Teller in the State of Colorado, to be in a state of insurrection and rebellion.

It seemed to be an instant replay of past events as Adjutant

General Sherman Bell arrived in the mining district at midnight the same day to take command. It appeared that the labor problem was moving to a rapid conclusion.

On the eighth of June, the offices and plant facilities of the *Victor Record,* the staunch supporter of the Federation, were ripped to pieces by the military, while the staff was arrested and thrown into the Bull-Pen. Ironically, the morning the plant was wrecked, the publisher ran an editorial calling on the Federation to call off the strike. The newspaper later presented a statement of damages — $4,206 — against the State of Colorado. Much to the chargrin of the union, when the damage bill was paid by the state, the newspaper suddently became an active supporter of that administration.

The next step in dealing with the union was deportation. On June 10th, seventy-two men were removed from the district under armed guard. They were loaded on railroad cars and shipped to a point near the Kansas border, where they were marched out of the state. One of the first to be deported was the former marshal of Victor, Mike O'Connell. In all, two hundred men were deported. However, many were warned, and they left with no further urging; and still others without warning could see what lay ahead and left of their own accord. The situation now resembled the 1899 Coeur d'Alene troubles.

The next move was to close down the mines. The military and civil authorities realized that the influence of the Federation could not be broken as long as the mines operated with union labor. The solution was to allow those miners to work who would renounce the Western Federation of Miners and take out mine-owner's cards. Those who refused were deported or forced out of the district.

Slowly, but surely, order and stability were restored to the Cripple Creek mining district. On July 26th, Governor Peabody issued a long-awaited proclamation declaring that "further application of military occupancy or authority is suspended." Governmental authority was turned over to civilian representatives, and the state troops were withdrawn. The bloody and costly strike was over. This was by no means the end of union activity, but for the time being Harry Orchard knew he had to keep moving.

CHAPTER 13.

Is this the hole in the wall dump?
. . . George Pettibone

With the law on his heels, Orchard and the Nevilles headed north out of Denver for Cheyenne, Wyoming. The trip by team took them four or five days. One of the interesting outcomes of the trip was the growth of a genuine attachment between the mass slayer and the teenager, Charley Neville. Orchard bought the boy candy, and in the evenings, while they were sitting around the warm and pleasant campfire, he played mumblypeg with him. The association developed into a fond and permanent relationship. After Orchard made his infamous confession and the trial of Haywood was approaching, the prosecution wanted Charley Neville as a possible witness, to help substantiate certain parts of Orchard's story. However, the killer displayed unusual concern for the boy's safety and sought protection for him against possible Federation reprisal.

The killer's obvious affection for Charley was an emotion he had never shown before, even towards his own daughter and his stepchildren. This was quite a different attitude from that of the man who had told Haywood he was going to desert his wife, Ida Tooney, because he was tired of supporting her boys, who were not even his sons.

Entering Cheyenne, the weary travelers lodged their horses in a livery stable and looked for Pat Moran's saloon. Moran was an old friend of Pettibone's and had once worked for him in Denver, selling clocks and lamps on the installment plan. After leaving Pettibone, Moran had opened a saloon in Rugby, Colorado, a mining camp, near Colorado Springs. He had moved again when a fight occurred and one of the participants died, going to Cheyenne in 1903 to enter the saloon business. Pettibone had given Orchard a secret identification pass word to be used when he contacted Moran, so the saloon keeper would know who sent him — "Is this the hole in the wall dump?" This somewhat childish phrase proved effective, and it probably showed Pettibone's interest in the legendary Butch Cassidy and his gang of outlaws.

Once identified as friends, Moran showed them a newspaper article which stated that both Orchard and Neville were wanted in connection with the dynamiting of the Independence depot. Startled, the killer read the physical description of himself and Neville, and suddenly realized they were after him. The article went on to say that the men in question were being sought in several states, including Wyoming.

John Neville was affected most by the news. He wanted to return to Cripple Creek and give himself up at once, but at the same time he assured his friend that nothing could happen to him. "They don't know anything on you, and I can be a good alibi witness for you," he told Orchard. The murderer's concern was not so much for the elder Neville as it was for his son, Charley. The young boy would be more susceptible to questioning and could innocently give everything away. Using every persuasive argument at his command, Orchard convinced Neville to continue the trip, reminding him of the arson job on his saloon. That did it. Neville finally agreed that the trip would go on as originally planned. The truth of the matter was that Sheriff Ed Bell in Cripple Creek was suspicious of Harry Orchard's connection with the mass killing, and about ten days after the fatal blast, a warrant was sworn out for his arrest.

By coincidence or by design, that same night Bill Davis and D.C. Copley arrived in Cheyenne on the late train. At one time Copley had been under indictiment for conspiracy to commit a felony in the blowing up of the transformer house of the Sun and Moon mine in Idaho Springs. He had met Orchard in 1903, in the library of the Cripple Creek Miners' Union No. 40. Copley was on his way to Butte, Montana, after he fled from Cripple Creek because the authorities were looking for him.

The men did not hesitate to hunt for Moran's saloon. There they confirmed the news that Cripple Creek authorities wanted Orchard and Neville. The pressure being exerted in the mining district against Federation members was unbelievable, the two men told Orchard. As he listened, Orchard knew he was justified in his decision to keep moving.

As they drank, Harry and Neville kicked around possibilities of relocating, toying with the idea of going into the saloon business. Neville had a good reputation as a saloon operator, and they had money. Orchard was certain he could obtain more from the Federation when he wanted it. To be successful they knew they had to set up in a town that had some life and yet was removed from the mining scene. Cody, Wyoming, they reasoned, was the place that met their requirements. Orchard knew that Neville would stick with him; the arson job they had pulled on Neville's saloon in Independence would insure that. And the move to Cody would give them both a chance to start over.

Another complication was added when Davis proposed that the killer forget his current plans and travel with him to the West Coast, at least until all the commotion died down. Orchard was not impressed with the idea. He was not overly fond of Davis, especially since he'd taken advantage of him in his early days at Cripple Creek. Davis was also quite a well-known figure in union circles because he'd been involved in a number of bloody strikes; in short, he was bad news. Orchard was also aware that the man had only about a hundred dollars in cash, barely enough to get him to the coast. And he would have no money to support himself once he arrived. So he refused to give Davis' proposal a second thought.

Orchard forced himself to attack the problems he and Neville faced if they were going to set themselves up in the saloon business. They readily admitted they needed more financing than the presently had. The Western Federation of Miners was the answer. Orchard had often bragged that he could get money from the organization anytime he wanted it. Now would be the test, but how could he send word to them? The solution seemed to lie with Moran, who was acquainted with Pettibone. Perhaps he could be persuaded to make the trip to Denver and to return with the necessary money. Pat Moran agreed to go for them since he had business there anyway. He caught the next train south.

Orchard sat down at the poker table and wrote a letter to George Pettibone, requesting five hundred dollars. The killer emphasized his request by saying he was considering going to San Diego to look at some property that Neville owned in that area. He was on firm ground with his lie since he had talked with Federation leaders about this proposition before. They had been receptive to the idea and had expressed a desire to obtain an appropriate hiding place and rendezvous for Federation men who were wanted by the law.

Orchard, Neville, Charley, and Davis retired to a room in the rear of the saloon where the bartender slept and waited. They wanted to be out of sight and also away from the sharp ears of the bartender. The bartender was a long-nosed, dark-complexioned man in his early thirties. A trait that set him apart from most bartenders was that he did not drink. This in itself was enough to make Orchard wary, and he was also puzzled by the man's behavior. Did the bartender know his identity? He preferred to stay away from him and hope his suspicions were wrong.

Time passed slowly that day — painfully slow. What else could one do in a saloon with time on his hands except drink, so drink they did. Orchard did one more thing — he shaved off his mustache. Surprisingly, to everyone except Orchard, Moran did return with a letter from Pettibone containing the five hundred dollars he had requested. Pettibone also wrote and suggested that Orchard seriously consider southern California and look for a place near the Mexican border. He suggested that it should be close enough to the border so that one could escape quickly into the sanctuary of the other country. He implied that Orchard should travel alone, to preserve the secrecy of the retreat. However, now that he had the money, Orchard decided to abide by his original travel plans. Now it became important to keep an eye on John Neville and on Charley. He liked John, but friendship could stretch only so far.

The party of three left Cheyenne and started west for Laramie. To others, they were simply eager sportsmen enjoying the great outdoors. Finding what looked to be an exceptional fishing spot on Crow Creek, they stopped to drown a few worms. The weather was ideal and the fishing good, so they decided to stay a couple of days. The two men had fortified themselves for the apparent pleasures by purchasing in Cheyenne, along with the necessary groceries, bottled beer, two jugs of liquor, and several boxes of cigars.

Davis and Moran rode into camp that afternoon. They said they came to socialize. While Moran fished the creek with young Neville, Davis got Orchard alone and told him that he wanted to go with them through Wyoming. Orchard was silent for a long time as he again considered the plan, but he still felt the same as before. He did not want Davis along, so he put him off by claiming they were already crowded and there was no room in the wagon for another man. Davis accepted the decision, but his disapproval was evident. That night Moran and Davis slept together while Orchard remained with the Nevilles. They would have two separate camps; that was the way it was going to be.

Davis and Moran returned to Cheyenne the next day, while Orchard and the Nevilles packed up and rode into Laramie where they took on more supplies. After a needed shave and a good meal, they were back on the road toward Casper, this time heading north. It was a leisurely excursion, no rush, no pushing, just riding in the wagon until they spotted a likely fishing area. They stopped, fished, and drank — as if they had not a care in the world. There was no scheming, no violence, no bloodletting.

The first thing Neville did when he entered Casper was to check the post office for mail. He had written to his wife from Cheyene and had told her to address any letters to Pat Moran in Cheyenne. Arrangements had been made with Moran to forward any corrspondence to Casper, but there was no mail. Strangely, Orchard did not object to Neville's receiving mail, certainly evidence that could be traced. Perhaps he did not wish to push his friend into any rash actions, and it would be natural for Neville to want the latest news from home.

Neville wrote to his family again, and this time he told them to address his mail to Cody. That night Neville and Charley were tired and wanted to get to bed early, but Orchard wanted to enjoy some nightlife. It was time to drink and gamble and participate in a few exciting activities. Knowing his weakness for poker and other games of chance and needing money to set up in Cody, the killer gave the five hundred dollars to Neville to hold for him until they left Casper.

After a few days, the party continued into Thermopolis. It was now about the 10th of July, 1904. The men set up camp near one of the hot springs and prepared to make the most of their situation. Since John Neville was breaking out with some type of skin disease on his face, possibly ring worm, he thought the hot baths might affect a cure. After nearly a week, Orchard showed signs of nervousness because of lack of activity. He wanted to ride into Cody to see what the prospects really were for the saloon business they had talked about. As Neville wanted to stay in Thermopolis a while longer and take the baths, it was agreed that Orchard would ride into Cody, look around, and then return to pick them up.

The killer rode northwest to Meeteetse, a small town also situated on the Grey Bull River. Not being particularly fond of travelling in the saddle, he decided to take the stage the rest of the way into Cody. He found the horse could also be sent back to the Nevilles in Thermopolis. In Cody, Orchard checked the post office and picked up mail. He received an important letter from Pettibone informing him that the law was hot on the trail and was eager to bring the bombing case to a close. The best course of action for Orchard was to keep moving. Orchard found a saloon, drew a beer, and then reached another decision.

Orchard managed to get John Neville by telephone and told him he had received letters which would be forwarded to Thermopolis, and that the horse was on its way back. Then he hesitated. How would he tell John that all the plans they had made would have to be changed? He decided to get straight to the point. Neville listened silently as Orchard informed him that things were not going well and that he would have to travel farther north into Montana country. On the other hand, Cody might be a good place to set up another saloon. It looked like a real going town, and Neville ought to check on it further. Orchard promised to write and keep him posted.

That night, Harry Orchard decided to have a last fling before he left Cody, a decision that proved disastrous. It all started out peacefully enough — a couple of belts of liquor, then the empty chair at the poker table. Lighting a cigar, he eyed his opponents through the blue haze. They looked like ordinary dudes, three markes. His assumptions seemed to be correct as he started winning. He filled the straights, drew the third king, and found he could bluff if he had enough money on the table. When he quit at nine o'clock the following morning, he was almost two hundred dollars ahead. He could ride out of town with some big money if he played the cards right. All he needed was a little sleep to clear his head.

Getting up that same afternoon, he showered, shaved, ate dinner and then looked up his poker playing friends. This time, however, his luck changed. Orchard was losing, and it was downhill all the way. The hundreds disappeared in short order, and he knew he could not make it at the poker table. A new game would change his luck, he decided. Orchard tried bucking Black Jack by betting fifty dollars a turn, frantically trying to recoup his losses. It was no use; he continued to lose steadily until he was broke.

Thinking fast, Harry Orchard went to the saloon proprietor, a Mr. Hall, and with all the gall of a born con artist said, "Lend me fifty dollars to get to Denver, and I'll pay you when I get back." Hall already knew from the man's constant boasting that he had a partner in Thermopolis, and that they intended to buy business property in Cody. So he gave the loser the money. After all, he was only loaning the man his own money, and it was no great loss.

It had been six weeks since Harry Orchard left Independence for the open spaces of Wyoming, and it felt good to him to be back in Denver. Renting a room under an assumed name, he met at night with Haywood and Pettibone at the store and reported on his trip. To account for his money, Orchard settled on the story of his investing in land with the eventual possibility of placing a saloon on his holdings. The reason for his return trip to Denver was to obtain more financial backing so he could work on the building. The tale seemed to satisfy the curiousity of the two men. Pettibone advanced more money, but after discussing the Cody situation carefully, Haywood and Pettibone felt that for Orchard to return to Wyoming right now would not be wise. They had heard from reliable sources that Cripple Creek law officers were moving in that direction, and the advice proved to be correct.

One of the current projects of the union leaders was to engage Arthur Baston to eliminate Governor Peabody, but Baston had floundered in his efforts so far to shoot the chief executive. Baston had been stalking the governor near his residence, but after questioning Adams and hearing the story of what he and Orchard had experienced in trying the same thing, he feared Peabody might have hired body guards.[48]

As for Harry Orchard, whose stock was high with union leaders and who was considered the number one boy when it came to murder, his talents were needed for a new job. The superintendent of the Highland Boy Mine in Bingham, Utah, Andy Mayberry, had discharged 150 union miners for taking part in labor demonstrations. The Federation leaders wanted to retaliate with drastic action, the kind that would show union men everywhere they had protection against that type of behavior.[49] They wanted to send Orchard and Baston to Utah to kill Mayberry, but the plan was shattered when Orchard told them that he had lived in Bingham and was personally acquainted with Mayberry. Haywood realized it would not be good business to have Harry Orchard recognized, so the murder mission was scrubbed. Apparently, he had little faith in the ability of Arthur Baston to pull the job off himself.

Despite the setback, Haywood would not let the topic of assassination drop. He wanted to show the power of the organization. He wanted to prove to the world that the Federation always punished its enemies. It was an easy matter to obtain money. The Federation had received more money the month after the Independence explosion than any month during the strike trouble. Haywood once estimated that the Independence depot explosion had been worth $30,000 to $60,000 in contributions.

An alternate plan had been to kill Frederick W. Bradley, manager of the Bunker Hill and Sullivan mine during the 1899 troubles in the Coeur d'Alenes. Bradley was presently heading the Mine Owners Association in California and was actively engaged in driving the Federation out of that state. Bradley was one of two people connected with Federation activities in Idaho who had been marked for death; the other man was the former Idaho governor, Frank Steunenberg.

Haywood had already sent Steve Adams and Ed Minster to California to murder Bradley, but they had failed. Perhaps Orchard could finish the job. He considered it a challenge. Enthusiastically, Orchard asked for a chance to make the kill. Ironically it had been Orchard who had helped destroy the great Bunker Hill and Sullivan concentrator in 1889, and now he was going to kill the former manager of that great mining company.

At this time, Steve Adams had also been given an assignment of merit. He was in Wardner, Idaho, responding to a summons from Jack Simpkins who wanted help in getting rid of claim jumpers who had made the fatal mistake of trying to grab Simpkins' land claim. After that job was completed, Adams was slated to go to Caldwell in southern Idaho and eliminate Frank Steunenberg. This would balance the Federation accounts in Idaho, for the intended killings were part of a systematic program designed to eradicate union enemies. If the killings were successful, they would alter the destiny of the organization known as the Western Federation of Miners.

So Pettibone purchased him new luggage, a train ticket, gave him $150 for expenses, and Harry Orchard was on his way to a new assignment, on the morning train for San Francisco. And as the Union Pacific made its way west to the "city of a hundred hills," he carefully plotted his next murder like a true professional.

CHAPTER 14.

Business is poor, Johnnie on the way.
. . . George Pettibone

No sooner had Orchard checked into the Golden West Hotel under the name of John Dempsey, than he began laying his plans. Using the city directory and telephone book, he located the home of Frederick W. Bradley and also his business office. Then he familiarized himself with those buildings and their surroundings by carefully studying the structures from the outside and by paying the utmost attention to the entrances and exits.

After this, he wanted to concentrate on the personal and work habits of his victim. Once he had established a routine or pattern of his prey's movements, he knew that a murder could be successfully accomplished. At this point, his scheme went askew. Boldly calling Bradley's office in an attempt to set him up for a murder try, he learned the man was in Alaska to inspect mining property and that he was not expected back for several months. Dammit! The very idea that the intended victim would not be at home had never crossed Orchard's mind. Haywood and Pettibone had never considered that possibility; they wanted the killer out of Denver and away from the attention of local authorities. Orchard was furious — he had come all the way to San Francisco for nothing!

Writing Pettibone, he informed him of the latest developments. The letter was written in a cipher code as a precaution in case the correspondence fell into the wrong hands. In it, the killer asked for a payment of one hundred dollars.

During the time Orchard was waiting for a reply from Pettibone, he read a newspaper account on the arrest of John Neville at Thermopolis, Wyoming, as a suspected accomplice in the mass slaying at the Independence depot on June 6, 1904. Neville was in the process of being transported back to Cripple Creek. The article went on to relate that the arrest of another fugitive, Harry Orchard, was expected to take place soon. Ironically, Orchard had escaped from the law in Wyoming by losing every cent he possessed in wanton gambling and had been forced to retreat from the state. What a fortunate loss! But Neville! What would happen to that poor fool now? Orchard brushed the thought aside and concentrated on his own predicament.

A telegram, confirming the news reports on Neville, arrived from Pettibone. It read: "Business is poor, Johnnie on the way." The telegram meant things looked bad, and that Orchard should look out for himself. He understood the intent of the message only too well. Shaken, Orchard checked out of the Golden West Hotel and rented a small, private room on the west side of town. He also started to limit his activities to the cover of darkness.

But there was another big issue to be resolved before the killer could hole up completely. He had not yet received money from Pettibone, and he desperately needed it. Still, he wanted to avoid the Golden West Hotel and, if he could, to avoid receiving the telegram himself. Orchard reasoned that if the telegram could be traced, the best way to trap a man was to wait for him to claim it in person. So he consulted the telephone book and located the name of the secretary of the local bartenders' union, Peter L. Huff. Orchard went to Huff's office at 211 Taylor Street and introduced himself as a union man from Colorado — a man with troubles. Whatever he told about his circumstances, it proved to be effective, as Huff agreed to help him in securing his mail.

Since any correspondence to Orchard would come to the Golden Gate Hotel, the secretary of the bartenders' union arranged to have all mail routed to Stockton and then returned to the post office on Mission Street. This was requested of postal authorities because Mr. Dempsey was a salesman on the road, and if he missed the delivery in Stockton, he could claim any mail when he again arrived in San Francisco.

Orchard, with Huff to identify him, picked up the letter he awaited on August 18, 1904, at the delivery window. It had been originally sent from Denver to John Dempsey at the Golden West Hotel on August 10, 1904. Pettibone's letter warned Orchard to lie low and be extremely careful. All letters were to be destroyed immediately. The requested money had also been enclosed.

Orchard was able to read Denver newspapers regularly, so he was aware of what was happening in Colorado. He also kept out of sight as much as possible by keeping his meals down to one or two a day and foregoing liquor and night life. For a man of Orchard's disposition, the routine soon became unbearable, and he found himself becoming suspicious of everybody and everything. He knew he had to get hold of himself before he fell completely apart. To do this, Orchard left the "City by the Golden Gate" and went to Caliente Springs in the southern part of the state where he played cards and rested for the better part of a month. Feeling more sure of himself, he returned to San Francisco and found a sleeping room near the Presidio.

Upon his return Orchard moved around more and found companionship associating with troopers stationed at Presidio. Booze, whores, and poker were a common denominator for friendship. Meanwhile, Pettibone sent money regularly, usually one hundred dollars.[50] In one instance, Pettibone prepared a letter for Orchard and enclosed "a union card and a masonic charm." A clerk working for the merchant, Jacob Wolff, took the letter to the post office and registered it. He noticed it was addressed to a "J. Dempsey."

Orchard soon discovered there were several areas in San Francisco where a wanted murderer could go and not be recognized. There he could fit in with the existing element and be right at home. One was Chinatown, and the other was the infamous Barbary Coast. Chinatown was a city in itself, and whatever pleasure a man sought could be found in that mysterious district. The Chinese loved to gamble, as did many of the visitors. The smoking of opium was limited for the most part to the Chinese, although some adventurous transients did occasionally indulge. Prostitutes were available from the renowned "four Bittee" cribs to the more elaborate establishments.

To the west of Chinatown was the Barbary Coast with the waterfront as its east boundary. Over the years the coast had calmed down, but a man on the "make" could still have a night to remember along Pacific Street. Liquor and girls were in abundance, and the numerous saloons and dance halls welcomed all with loud, brassy band music, steam pianos, and gramophones. Few cared what profession or reputation a man had; the important thing was how much money he had. This was the Barbary Coast, and men of Harry Orchard's class and stature reveled in it.

Early in September 1904, Orchard began planning for Bradley's return. He toyed with the idea of again using dynamite for the kill. He was fascinated with the mechanics of a bomb and went so far as to purchase ten pounds of gelatin powder from the Judson Power Company on Market Street, explaining to the clerk that he needed the explosive to blow up some stumps. The clerk recorded the buyer's description and his assumed name, and Orchard picked up the powder and caps at the company warehouse on Third and Townsend Streets.

At the drugstore, he tried to purchase some sulphuric acid and chloride of potash, but the druggist questioned the sale and inquired its use. "Experimenting," Harry told him. Not satisfied, his next question was, "What kind of experimenting?" The hot-headed murderer shot back, "None of your business." Needless to say, no sale was made. The next druggist offered no such obstacles, and the sugar was easily stolen from a cafe. The killer was now ready to continue his mission of death.

Bradley's apartment house was at 1404 Washington Street, on the corner of Leavenworth and Washington Streets. It was a large, three-story building which housed six families. When Orchard was looking for an appropriate room, he dropped in at a neighborhood saloon and store for a drink and met the proprietor, Laurence Giubinni. During the conversation he mentioned, "I want a room which overlooks Fred Bradley's place of residence." Giubinni could attach no special meaning to the statement. With the proprietor's assistance, he met the hired help of the Bradley household. Orchard arranged to be in the store when the women shopped for food and actually managed to deliver groceries to the victim's kitchen on several occasions.

One previous time, he had knocked on the door of the Bradley apartment to assess for himself the exact suite of rooms. The door was answered by one of the servants he had met. Taken by surprise, Orchard told her he wanted to see Mr. Bradley. Without hesitation he was led into the sitting room where he met his victim's wife. She informed him her husband was out of town, but she gave Bradley's expected time of return. The incident was forgotten and was never related to future happenings.

Meanwhile, Orchard secured a sleeping room at 1326 Washington, a few doors away from the Bradley residence. Since the structure was built on higher ground, his room not only faced the victim's residence, but also was on a level with the Bradley apartment. He could actually observe movement in the rooms. It appears that the assassin had chosen his accommodation with care, using professional skill in positioning himself for the kill. Harry Orchard rented the room on Sunday and introduced himself to the landlady, Florence F. Soward, as Mr. Berry. Mrs. Soward noticed that he was dressed like a working man, a common, undistinguished sort. The furnished room cost him two dollars and fifty cents a week, and he was more than satisfied.

The landlady soon observed that Mr. Berry was an odd one.He kept late hours, coming in early in the morning and usually slept or stayed in his room until eleven or even one o'clock. In sweeping his room one day, Mrs. Soward was going to move his suitcase from one position to another, but it was too heavy for her. The heavy piece of luggage caused Mrs. Soward to speculate what was inside that weighed so much. The woman did not realize that the luggage she was attempting to move was an instrument of death and was full of explosives and all the component parts of a bomb.

The Giubini grocery store and saloon was situated on the corner opposite the Bradley residence. It proved to be useful to the killer. First, the saloon was a good place to drink, and second, the girls employed by the Bradley family shopped at the grocery store. They unknowingly became a source of information for the man planning death. From them he learned intimate details about the Bradley family which could prove invaluable later on.

When the victim returned to the city the latter part of October, Orchard was ready to formulate his final design for Bradley's extinction. Even though he had the explosives to do the job, the killer was hesitant about using them, as he could not isolate a situation where he could properly use them. He decided to use the shotgun in much the same manner that he had done in the

Gregory killing — stalk the victim, catch him unaware at night, and blow him apart with short range blasts from the shotgun.

The assassin followed a routine that was reminiscent of other days while he stationed himself across the street from the Bradley's apartment with his weapon and waited for the victim to return at night. This time, the days dragged on without presenting an opportunity to finish the job. Orchard blamed the street lights for there was not enough light to distinguish Bradley from the rest of the building's inhabitants. What made the situation even more difficult was that the entrance to the building was through a large archway, and from there the people would go to their own individual apartments. Trying to spot an unfamiliar person at night, under poor lighting conditions, was almost impossible, he quickly discovered.

The true reason for his failure might rest with the fact that the only man he had ever executed face to face was Gregory, and that had happened with assistance and encouragement beside him. Orchard had blown every chance to kill his victims on a personal encounter basis, even when he was the only armed individual on the scene. The excuse of poor lighting was a good rationale for his own failure to kill in this manner. He was much more successful when he used other methods.

Meanwhile, the frustration in being temporarily thwarted in getting rid of Bradley began to torment Orchard. The months of waiting for the victim to return and not being able to finish the job added to his growing despair. To make matters worse, Pettibone had refused his last request for money, and the killer was getting desperately low on finances. In a fit of depression, he told himself that he was being taken advantage of, that the union was through with him. There were not going to be any more pay days because they thought he would not dare go back to Denver because of the warrant of arrest issued against him. The more he thought about it, the more he wanted to return and confront Pettibone and the others. He knew they wouldn't dare to refuse him.

Unknown to Orchard, the reason for Pettibone's refusal to send money was that John Neville was trying to "shake down" the union leaders. Neville had been in jail for several months and then released for lack of evidence. He and his son returned to

Thermopolis, Wyoming, to claim their horses and wagon and headed back to Colorado. By this time Neville had thought up his own scheme for making money, without working. Had his association with Orchard led him to try for an easier life? The first step had been the burning down of the saloon for insurance money, and now a little blackmail racket loomed.

In Denver, Neville forced a meeting with Haywood and Moyer, telling them point blank that he knew all about their operations, particularly the details concerning the Independence depot explosion. For emphasis, Neville completely lost his head and mentioned that his good friend, Harry Orchard, had told him the whole story. Since they knew he had spent a great deal of time with Orchard immediately after the Independence depot affair, they decided he just might be telling the truth. Neville demanded money to keep his mouth shut. Moyer grew especially angry because of his threatening, but he quieted down enough to discuss it logically with Haywood.

The best approach, they reasoned, was to counter Neville's blackmail with a little of their own. If Neville kept on talking like that, the police might be just as interested as they were in finding out about the mysterious fire that had destroyed his saloon. That information ought to be good for a year or two in prison. The ruse worked, and John Neville backed down. He knew he was beaten.

The fear of having to go to prison himself was too much to chance; John Neville pulled up stakes and left Colorado, but he was a marked man as far as the "Inner Circle" of the Western Federation of Miners was concerned. By challenging them, he had signed his own death warrant.

CHAPTER 15.

I had a little powder of strychnine made for each bottle; raised the paper cover and emptied one of these in each bottle of the milk and cream, and stirred it up a little, and pressed the paper cover back again; and left and went back to my room.
. . . Harry Orchard

In San Francisco, Orchard was reaching the breaking point. He was furious at his employers, and the killer instinct demanded blood. First he would have Bradley's, and then he would return to Denver for a showdown. Like an enraged bull, he again reviewed the ways he could use to kill the mining executive. The shotgun bit was taking too long and was not at all to his liking. Maybe an entirely different technique was needed, something he had never used before, something like poison, perhaps. A poison could be put secretly into a liquid substance and then be taken orally, without the murderer even being around when it was administered. This was the type of approach that appealed to a man of Orchard's temperament. The poison could be put into a bottle of milk, for instance, and nobody would be the wiser.

Milk was delivered to the Bradley apartment every morning; this Orchard learned through his contacts with the hired help. If the poison could be put into the milk bottles immediately after they were delivered and before they were taken into the apartment, the plan might work. At any rate, it was worth a try. Orchard put his plan into operation by purchasing some powdered strychnine. He knew very little about the bitter and poisonous drug except that a little went a long way. As he examined the colorless crystals, they seemed harmless enough. Yet this unimpressive looking drug caused death by effecting a difficulty in swallowing, then a sudden backward bending of the body, and finally a twitching that developed into uncontrollable convulsion. It was a tough way to go, but what the hell — a job was a job.

On that particular morning, Ed Adair of the Guadalupe Dairy made his delivery of a one quart bottle of milk, a one pint bottle, and a half-pint of cream. The time was about six-thirty a.m. Harry

Orchard recorded his actions as follows:

> I had a little powder of strychnine made for each bottle; raised the paper cover and emptied one of these in each bottle of the milk and cream, and stirred it up a little, and pressed the cover back again; and left and went back to my room.

The psychotic nature of the murderer again surfaced when he was perfectly willing not only to poison Bradley, but also his wife, his child, and the domestic help to achieve his objective. He slipped into bed and slept soundly. Orchard always seemed extremely tired after any hazardous action.

About an hour after the milk delivery was made, the Guadalupe Dairy received a telephone call from an angry, dissatisfied customer, a Mr. Frederick Bradley of 1404 Washington Street. He asked that the dairy discontinue his milk delivery, as he thought the milk that was delivered this morning was poisoned. Bradley commented on the incident as follows:

> The attempt to poison me with strychnine in the milk was frustrated by my habit of early rising, my custom being to breakfast before anyone in the house. On this morning I noticed a bitter taste in the milk, spit it out, and had the rest put out of the way. Several innocent lives were thus saved. The milk had only been delivered a few minutes before my tasting it.

As soon as possible, Mr. Oliver Crook, manager, hurried to Bradley's apartment and received from Mrs. Bradley a quart bottle of milk. He took a small amount of the questionable liquid and found it tasted as if it were full of quinine or some other bitter stuff. The bottle of milk was then taken to Dr. A.C. Bothe, city chemist. Dr. Bothe's analysis indicated there was enough strychnine in the milk to kill an army regiment.[51] Unaware of what had happened, Harry Orchard rose about ten o'clock, dressed and shaved, and as usual sauntered over to Guibinni's saloon where he drank his own liquid breakfast and read the morning newspaper. Today, however, he noticed a bottle of milk on the bar and had a strange feeling that particular bottle was one of those into which he emptied the poison. When he innocently asked the saloon proprietor what a bottle of milk was doing in a saloon, Guibinni explained that the domestic help from the Bradley apartment across the street had brought the milk over for him to taste. He had tasted it and had found it bitter. As a matter of fact, he could taste the bitterness in his throat even now — something was definitely wrong with it.

Thwarted a second time, Harry Orchard, professional assassin, decided to resort to the method that had brought him success in the past. It was time to try the familiar dynamite routine, for he had become an expert when it came to making a bomb for human destruction:

> I bought a piece of five-inch lead pipe about a foot long at a plumber's and put wooden ends in it. Then I hammered one side of it flat, so it would lie straight without turning over and I cut a piece out of the other side and turned back the flap, and fastened a little vial on this so that when you filled it with sulphuric acid, and you pulled out the cork, the acid would run out into the hole in the pipe. Then I filled up the lead pipe with about five or six pounds of no. 1 gelatin and put some caps and sugar and potash on top of this and opposite the hole in the lead pipe, so the acid would fall on them.

The ingenious device constructed, the only remaining detail was to find a method to start the explosive action. Orchard solved this by developing a pull mechanism. A cord was attached to the cork stopper of the sulphuric acid vial, and the other end of the cord was to be secured to a screw-eye mounted in the door. When the door was opened, it would pull the cork from the vial and release the acid to detonate the bomb. In short, the explosive package became a booby trap which would make the victim blow himself up.

The killer then experimented to determine the exact amount of force it would take to pull the cork out of the vial without disturbing the bomb in anyway. The closet door in his sleeping room became an experimental station. A screw-eye was fastened

solidly in the door, then a cord was attached to it and to the screw-eye secured in the cork of the acid vial. The bomb at this time contained dynamite, but neither the blasting caps nor the acid in the vial. Finally, Orchard regulated the device so that it took very little force to remove the cork, and yet it did not disturb the bomb to any degree. The bomb casing was designed to lie flat, anchored to a certain extent by its own weight.

In cleaning Orchard's room during this period of experimentation, Mrs. Soward came across shavings of lead and wood. She also found fastened to the closet door a hook about twelve inches from the floor, and to the hook was tied a cord. The other end of the cord was joined to a small cork. She cleaned around this odd contraption, wondering what it was used for. In straightening the dresser bureau, Mrs. Soward came across three bottles filled with some kind of liquid. Mrs. Soward assumed her roomer must be an inventor of some sort, and concluded her speculations at that point.

By the early part of November, 1904, Orchard had definitely decided upon using explosives to eliminate Bradley. The killer had perfected his means of destruction, which needed only the addition of caps and acid to be operational. The only remaining task left that would make the bomb totally effective was the correct positioning of it so the intended victim would receive the full effect of its contents or at least enough to cause death. Now the problem narrowed down to a where and when situation. After much deliberation, Orchard decided the main door opening into the Bradley's apartment building would be the one. The only question left was, when?

By constantly watching the apartment house, Orchard knew that Frederick Bradley was usually the first person to leave the building in the morning. To determine his victim's movements for a short period of time the killer telephoned him on the night of November 15, at his apartment. Using a fictitious name, Orchard said he owned mining property in Goldfield, Nevada, a current boom town, and that he wanted to raise money or at least get someone with money interested in developing the property. Orchard then asked for an appointment to meet with Bradley to discuss the situation. Bradley told him that he would be most pleased and suggested they meet the next morning at his office. Orchard thought quickly. He was quite ready to use the bomb, but he would need another day. To buy time, he asked Bradley if it were possible to make the date for the following morning, November 17th, at nine o'clock. Bradley readily agreed, and the appointment was made.

On the night of the sixteenth, Orchard boldly walked to the front door of the Bradley apartment building, and after first checking to see if anyone were around, he fastened a small screw-eye to the outside door. All that was needed now was the actual attachment of the bomb itself, which would be accomplished early the next morning. The killer then went back to his own residence, found his landlady, paid his rent, and casually mentioned he would be leaving for Idaho in the near future. Still later that same evening, Orchard packed and took his luggage down town and left it at a saloon where he intended to pick it up the next day. The only items he had with him in his sleeping room were the clothes he was wearing, a pistol, and of course, the bomb. The last thing he did that night was to take the container of death and wrap it up in a newspaper to disguise it as much as possible.

The next morning, the seventeenth, about 6:30, Harry Orchard took the bomb and calmly proceeded to the main door of the Bradley apartment house. He laid his parcel down and hooked a cord extending from it to the screw-eye that was mounted on the door. when he placed the doormat over the bomb, only the cord protruded, and it could not be detected from any distance. The killer finished just at the right moment and left to catch a cable car for the downtown area, not bothering to look back.

About 7:40 a.m. Frederick Bradley finished his breakfast and morning newspaper, lit his cigar, and started downstairs for the front door. He had the cigar securely placed in his mouth and some papers in his left hand. Bradley reached the foot of the stairs and walked toward the front door, his right hand reaching for the doorknob. He opened it and, "A light flashed from the end of my cigar, a deafening explosion followed, and I was smashed to the ground." He lay on the ground covered with plaster and other debris, bleeding from the neck up, deaf and blind, but miraculously he was still alive. He was seriously injured, but he had survived.

The vestibule floor of heavy white tile laid in concrete and the marble steps were blown to pieces. A fireman on duty several blocks down the street heard the explosion report. Several years later, he compared it to the dynamiting of buildings during the fiery holocaust that followed the city's earthquake, and he claimed that it was as loud as any he heard during that period.

Meanwhile, Orchard was in the downtown area in the process of renting a sleeping room. Exhausted, he felt the pressure of the last few days overwhelming him. He threw himself on the sagging bed without knowing what had happened and slept soundly. This time, he felt, his plan had to work.

A while later, Orchard was awakened by a loud banging noise on the door. Suddenly apprehensive, he asked what the intruder wanted, but he was told to open the door and find out. He retorted, "Get the hell away and leave me alone." A short time later, the knocking resumed. This time a voice informed him it was the sheriff, and he had better open the door. Now instantly awake, Orchard grabbed his pistol and told the sheriff to wait until he was dressed. Sick with anxiety, his one thought was that he had been seen leaving the bomb and had been followed. Hurriedly dressing and with his pistol in his right hand, he opened the door a few inches. Gathering courage, he prepared himself to shoot his way out, if necessary. Orchard did not want to put himself in the same position that he had been caught in with Moyer in Ouray.

Peering out of the opened door, he saw only the landlady. then the killer listened in amazement as the woman explained that the sheriff had come to seize her furniture and was in the process of removing it. He could scarcely believe what he had heard. The coincidence was absolutely stunning. Immensely relieved, Orchard walked out of the building. It was close to four in the afternoon, and he was a free man.

As the killer walked briskly down the street, he stopped to pick up an afternoon newspaper and scanned the pages, but found no mention of Bradley. There was no mention of the explosion. Suddenly, the familiar queasy feeling returned. What in the hell had gone wrong this time, he asked himself, with mounting panic?

Flashing across his mind was the realization that he had left the practice screw-eye in the closet door at Mrs. Soward's. That was not the brightest thing in the world to do. Harry Orchard was full of doubts. He wished he could leave the city, but he couldn't get very far. He desperately needed money.

Walking on the west side of town, he left the busy part of the city. Hungry, he went into a restaurant. Picking up a late copy of the *Evening Post*, he seated himself and took a sip of the hot, strong coffee. And there it was — a picture of the explosion scene and a full account of the action. Dammit! Bradley was still alive, but he was hurt badly. But it was better than nothing at all, and Haywood and Pettibone would be pleased. Now he could get some money for the job. It was a good day, after all.

One unusual twist to the attempted murder of Frederick Bradley was that the cause of the explosion was said to be a gas leak within the apartment house. It was supposed that Bradley's lighted cigar had ignited the escaping gas and it had exploded. This theory brought an even stranger ending to the affair. Walter H. Linforth, an attorney and the owner of the building, brought suit against the San Francisco Gas and Electric Company, claiming that the explosion was the result of a gas leak. He won the case and was awarded damages.

Ironically, Bradley himself believed, publicly at least, that the explosion was caused by escaping gas. The mining man declared he had smelled only gas fumes, no powder fumes, an odor with which he was most familiar. It has been suggested that Bradley knew very well that he had been blown up by dynamite, but, since his wife had feared just such an incident, he supported the leaking gas theory. Bradley is reported to have told a personal friend, Albert Burch, also a mining executive and former superintendent of the Bunker Hill and Sullivan mine, that the force of the bomb threw him back into the house; this burst a gas meter, and the explosion of the gas blew him out into the street.

Linforth testified that tenants complained of gas fumes prior to the explosion. The building's tenants used gas grates, and when the atmosphere was damp, these grates threw out an odor similar to that of escaping gas. A few days prior to the explosion, the gas company had sent an expert to make an examination of the premises because of the complaints, but he had found no gas leakage. There was also some question whether a burning cigar could ignite that particular kind of manufactured gas. Other gas companies, handling the same manufactured gas, had no rule against smoking because they believed the gas could not be ignited in that manner. And so the investigation that Orchard feared would follow the explosion never materilized. The killer was saved by the acceptance of the escaping gas theory. Mrs. Soward did find incriminating evidence in his room — the metal wood shavings, the screw-eye attachment on the door, and the strange behavior of Mr. Berry — but she reported nothing. Triumphantly, Orchard sent Pettibone a copy of the *Evening Post* containing the story and told him to send more money at once, which he did. He mailed it to Harry Green, a name Orchard was using at the insistence of Pettibone who claimed it was a "good name for anybody."

Now he waited for Bradley to die, but no obituary appeared. About ten days after the explosion, Orchard decided to check personally on the victim. Mustering all his gall, he telephoned Mrs. Bradley and introduced himself as Herb George, an organizer for the Citizens' Alliance. He claimed he was calling to inquire concerning Mr. Bradley's condition and to see if there was anything he could do to help. Mrs. Bradley told him that her husband's injuries were still of a serious nature. Although he was out of danger, his eyesight was still a question as they had not removed the bandages yet. "Mr. George" offered his regrets and terminated the conversation.

Financially sound again, thanks to George Pettibone, and feeling that he could move about the city without fear of detection, the killer emerged from hiding. Once more he drank and gambled with the troops stationed at Presidio. There was nothing more satisfying to him than sitting around a table with a group of men and playing poker. Harry Orchard was living the good life again, convinced that he had most certainly earned the right to indulge himself once more.

On December 4, 1904, Orchard read an announcement that D.C. Copley was giving a stereoscopical lecture at the Alhambra Theater on the corner of Eddy and Jones, dealing with the Cripple Creek troubles. The proceeds of the lecture were to go to the men on strike. That night, Harry turned up at the lecture and surprised the speaker. The last time Copley had seen Orchard was in Cheyenne, and much had happened since then. They talked a short time before Copley excused himself to start the lecture, giving the killer his address of 242 Kearney Street and agreeing to meet at another time.

A week later at six o'clock p.m., Orchard called on Copley at the latter's room. Small talk came quickly, and Orchard said he was making easy money gambling with the soldiers at Presidio. Then he brought up the matter of Bradley being blown up, adding that the mining executive had gotten what he deserved. Getting started on the subject of violence, the murderer speculated that more men would "get it" later. One in particular was Frank Steunenberg of Idaho, the son-of-a-bitch who had made him sacrifice his Coeur d'Alene mining property. He, Harry Orchard, would be a wealthy man now if that bastard hadn't called troops into the mining district. Death was too good for men like Steunenberg, he speculated.

Orchard had told the Steunenberg story so frequently that he was beginning to believe it himself. In all probability, he had started peddling the Steunenberg story to get attention and sympathy, and to let people know he could have been a wealthy man. It was better to blame Steunenberg for losing his interest in the Hercules mine than to tell the truth — that he had lost it at the poker table. Later, he began using the former governor as a scapegoat, an object that he could pour his hatred into, so as to drive himself on to kill and then justify his own horrifying actions.

On another visit to Copley, the murderer talked wildly about the possibility of taking twenty dollar gold pieces, drilling a hole in them, and then filling the coin with base metal and passing the money off as legal tender. Such schemes showed that Orchard was restless and wanted to return to Denver where the real action was. What he needed was a disguise in which to travel, to prevent identification by anyone who might still be investigating the Bradley explosion. Copley, now a pal, called on Orchard and found him changing into a soldier's uniform. He had obtained his pants and hat from a military tailor and the coat from a soldier with whom he was acquainted. Not satisfied with just wearing a uniform, he thought he needed a pair of glasses to complete the masquerade. Copley obliged him by buying a pair.

In early December, a soldier in uniform boarded the eastbound Southern Pacific train for Denver. He was leaving the "City by the Golden Gate" for the last time. That beautiful city that the murderer knew would be reduced to an ash heap in April, 1906, destroyed by an earthquake and fire. Albert E. Horsley, alias Harry Orchard, John Dempsey, Harry Green, Herb George, and Mr. Berry, was leaving for good. What a time he had had in San Francisco! Now, he was going back to Denver as the number one assassin of the Federation, eager for a new job to maintain his reputation. If he knew Moyer and Haywood, they would have one for him soon .

CHAPTER 16.

When they are through with you, they will treat you the same way.
. . . Steve Adams

Harry Orchard was in high spirits when he arrived in Denver near the first part of December, 1904. He rented a sleeping room on California Street, not far from George Pettibone's store, and once he was settled in his new surroundings, to celebrate his success, he telephoned his good friend Pettibone and asked him to come over for a little talk.

Steve Adams and the one-time justice of the peace showed up almost simultaneously. Adams had returned in September after helping union leader Jack Simpkins kill two claim jumpers on Marble Creek in North Idaho, and he was just as interested in the exploits of his murderous friend as was Pettibone. Besides, Adams had a little boasting of his own to do. To the satisfaction of his listeners, Orchard related the San Francisco adventure in detail, all except one small item. For some reason he left out the poison episode. Whether this was intentional or inadvertent remains unknown.

When they were informed that Bradley was still alive, although probably permanently disabled with sight and hearing defects, Pettibone seemed pleased, but he expressed real regret that Bradley had not been killed outright. Steve Adams, as a professional assassin himself, probably felt some consolation in his partner's failure to kill Bradley, since he, too, had made a trip to California in 1903 to eliminate the mining executive and had also failed to accomplish his mission.

Orchard had a few questions of his own for Pettibone, for he had waited a long time with no pennies in his pocket. Where in the hell was the money he had asked for? And what did Pettibone mean by asking him to call off the Bradley kill after all the work he had gone through to set it up? On the defensive, Pettibone explained in detail the blackmailing attempt made by Neville and what a shattering experience it had been. At that particular time, the union officials had steered away from any violence that could be laid at the door of the Federation and might provide Neville with further material to support his assumptions. They had been running scared, but now things were getting back to normal. Neville had been chased out of the state, and at the last report he was heading for the new Nevada bonanza, Goldfield.

Orchard saw Haywood in a day or two at Pettibone's store and again basked in the big man's presence. Haywood inflated the killer's ego even more when he mentioned that it was better in this case that Bradley was not killed, but was only permanently disabled. That way "he would be a living example" of Federation revenge. The union leader even suggested he might write Bradley and tell him how the whole thing had happened — that would certainly give him something to think about!

Steve Adams and his wife were living in Denver proper on Clarkston Street. They invited their good friend, Harry Orchard, to move in with them. After a few weeks in his small, confining hotel room, he gladly accepted. Several Cripple Creek comrades had shared the Adams' hospitality since his return in September. Bill Easterly had spent some time with him, and Billy Aikman was living there when Orchard moved in. Aikman had spent a week in the Bull Pen after the Independence depot explosion, and because of his reputation he found it safer to leave the hostile mining district while he could. Aikman had found a job as a city street cleaner and was contributing a little money to the household budget. Orchard was receiving money from the paymaster, Pettibone, although it was not a large amount. In any event, these arrangements kept Haywood from being directly involved in the financial transaction. And as future events would show, it proved to be a life saver for the Federation leaders.

On January 2, 1905, a group of prominent labor leaders met in Chicago to begin laying the foundation for a new and revolutionary industrial union. Among those present were three representatives from the Western Federation of Miners — Charles Moyer, William Haywood, and John O'Neil. O'Neil was the fiery editor of the *Miner's Magazine*, the official printed word of the Federation. The January conference went strong for three days, and its results were expressed in a manifesto which spoke of labor's grievances and suggested a movement that would correct these conditions. In essence, it proposed "one great industrial union founded on the class struggle."

Haywood drafted the manifesto, and it was evident that his ability would carry him far from the regional affairs of the Rocky Mountains mining camps into national spheres of influence. He also made it evident that he was concerned about the plight of the unskilled workers and that he wanted to include them in this new organization. Big Bill Haywood proved he was a man to be reckoned with while Moyer, resentment mounting, stood in the shadows. The short conference was declared to be a success, and a follow-up meeting was planned for June, 1905, in Chicago's Brands Hall, to organize this new revolutionary working-men's union.

After his return, Haywood discussed his visions of the new organization that would be called the Industrial Workers of the World and remarked to his number one assassin:

> This order is something like a wagon. The officers are the hubs, and between each spoke is a union embracing all crafts or professions, even to bootblacks. The foundation on which the hub turns is the emergency fund.

Excited, Haywood went on to explain that when the order was perfected and had all kinds of workers — men and women, and several hundred thousand dollars in the coveted emergency fund — there would no longer be any strikes. If demands were made of an employer, and he refused to concede to them, they would simply remove him. And so there would be no more strikes and no strike benefits to pay out, a provision that was presently weighing heavily on the Federation finances.

As far as positions in that organization were concerned, Haywood wanted to be secretary-treasurer or nothing at all. At no time in his life was this dynamic individual more of a radical-activist than during this violent period. If he were president, he would not be able to handle large sums of money, and for Haywood to accomplish what he thought was essential, financial backing was imperative. When he wanted an opponent of labor removed, he wanted to have the power to pick a good man and say, "Here is $3,000. Get this man out of the way. Then take this money, buy a farm, and settle down, and be comfortable for the rest of your life."

Orchard listened intently. He liked that part about having enough money to "buy a farm, and settle down, and be comfortable for the rest of your life." For enough money, he would do anything Haywood suggested.

Elated by the success of the San Francisco bombing, Haywood and Moyer selected their next victim — Judge William H. Gabbert, Chief Justice of the Colorado Supreme Court. That court had decided against Moyer at the time he was brought to Denver from Telluride on a writ of habeas corpus after having been arrested by the state militia. Moyer marked Gabbert for death, saying, "Get him! I want him out of the way."

Given the nod, Orchard and Steve Adams went after the judge. The killers, as usual, began the stake-out at the judge's private residence, where they knew that he began and ended the day. Adams was for staging the murder in the house or at least by firing into the house, as he had done when he had killed the unfortunate mine manager, Arthur Collins, in 1902. The killers, however, claimed they were unable to catch their victim in a favorable circumstance, even though the judge habitually walked back and forth daily from his home to the state capitol building. The biggest reason for the failure to assassinate quickly was a basic weakness of human nature: the cold winter weather of Colorado, and plenty of food and liquor without working for it. And so the assassination attempt bogged down in the cold slush of winter.

Also marked as a potential victim was Frank Hearne of the Colorado Fuel and Iron Company. That company had provoked the wrath of the Federation by pressuring the Colorado legislature in 1903, against passing the eight hour workday for the mining industry. Hearne had been spared sure death one night because the killers thought they couldn't make a fast enough get-away. David Moffat, a power in mining and banking circles, was also a candidate for elimination, but apparently no serious attempt was ever made on his life.

At this point, their attention shifted to Governor Peabody who presented a more pressing problem. He was the man who had incurred their enmity to a greater degree than any other in Colorado, and the final solution seemed to be his death. Orchard and Adams were only too eager to go to work again, and Peabody's elimination was a distinct challenge for their talents.

The death watch began with the usual shotgun routine. On various occasions, the murderers lay in wait for the Peabody carriage and attempted to ambush it, but cold weather constantly defeated them, and the numbed men once more retired to warmth and whiskey. Then, deciding to take advantage of new fallen snow, Orchard and Adams enlisted Billy Aikman and decided to use the familiar dynamite bomb. The Peabody family resided at 1128 Grant Avenue. Between nine and ten o'clock of every workday morning, the Governor walked up Grant Avenue to his office at the capitol building.

The bomb was finished near the stroke of twelve that January night, intended for use the next morning. It measured fourteen inches in length and eight inches in diameter. Constructed of lead and wood, it contained twenty-five pounds of dynamite, which Aikman had gotten from Idaho Springs, and was to be triggered by a pull action which would spill acid on the blasting caps and detonate the powder. It was placed in the gutter by the side of the walk on Grant Avenue and covered with snow at about eight o'clock, early for most businessmen. The decision to place it in the morning rather than at night was based on the cold winter temperature and the fear that the bomb might freeze and not function properly. A friend, Max Malich, hired a livery rig, and with Aikman driving, the rig was parked several blocks away. After securing the bomb, a soft wire was hooked to it and strung to Logan Street on Thirteenth Avenue, a block away.

Finally, the murderers pronounced the set-up foolproof and waited for their victim. The unsuspecting Governor left his home as scheduled and proceeded down the avenue. As he approached the lethal bomb, Orchard and Adams held the wire, in much the same manner as they had when they had destroyed the Independence depot. As Peabody approached, their hands tightened around the wire — it was now a matter of seconds!

Suddenly, a large coal wagon appeared out of the alley, its approach muted by the soft snow. Then another wagon wheeled into the area, and a crowd of people was now walking near the bomb site. The wire was relaxed, but when the unexpected traffic had cleared, the Governor was too far down the street to risk an explosion. Disgruntled, the killers dug up the bomb, wrapped it in burlap and carried it to the waiting rig, then drove downtown, returning later to pick up the wire.

Again they sought relief from the tensions and disappointment of the morning in a bottle of whiskey. Cursing, they poured their drinks. Adams had already been half shot when they left that morning; it was a bad thing to drink on a job that required iron nerves and a steady hand and mind. An angry Haywood blustered his idea on the project. "Why don't you plant the god-damned bomb in Peabody's desk and attach it to the rolltop?" But nothing ever came of that suggestion.

When a dispute resulted over the 1905 governor's election because Alva Adams, a Democrat, defeated Republican James Peabody, the arch foe of unionism, the Republican party claimed fraud and demanded an investigation. Haywood thought it best to lay off Peabody for the time being as it might hurt Adams' chances of being seated if anything happened. Adams was declared elected, but the legislature managed to unseat him with the understanding that Peabody would be sworn in and then resign in favor of the Republican lieutenant-governor. In this manner, Jesse McDonald became the sixteenth governor of the state of Colorado. At this point, Haywood decided that Peabody could be eliminated at their convenience.

After the abortive attempt on Peabody's life, an incident occurred which precipitated a crisis for the murderers. Aikman got himself arrested on a minor offense and gave the police his address. Afraid the police might follow up with an investigation, Adams and Orchard decided to relocate in a hurry. At Max Malich's suggestion they moved to an unfurnished five-room house near Globeville, a smelter area on the outskirts of Denver. Pettibone helped furnish the house by giving them some secondhand furniture which had belonged to Moyer. The household still consisted of four people — Orchard, who used the name of Tom Hogan, Adams, known as Dickson, his wife Annie, and her young son.

Money apparently was not too plentiful for the two men, or perhaps they just had a hard time holding on to it. Malich said they were "mighty hard up" and bought the cheapest groceries they could find. "Why," he explained, "they drank 25¢ coffee all the time." He lent money to Orchard five or six times and was always repaid whenever the killer "made a raise" at the gambling tables. On one occasion, Adams and Orchard were desperate enough to steal a sheep from the stockyards and slaughter it in Malich's saloon.

Malich claimed Orchard had come to see him after the murder of Lyte Gregory with a proposal: "Max, if I'm arrested for this, you're to remember I was out at your place playing cards." Presumably, at the proper time he would have provided the alibi in court if necessary. Orchard offered to return the favor by suggesting he would "do up" a William McDonald for a bargain price of one hundred dollars. McDonald was proprietor of the company store, the Globe Mercantile Company, which was in competition with Malich. The suggestion, however, never materialized, but Malich was impressed enough to offer "to bank" the killer if he wanted to go into the saloon business.

Max Malich, an Austrian, was a big man in the affairs of Globeville and was known as the "Boss of Globeville." Operating a grocery store and saloon, he was a leader among the Austrian smelter workers. Since he had been a smelter worker himself and spoke the same language, he was in a position of authority. Malich had been the mayor and councilman in the community and a strong supporter and member of the Federation. He had met both Moyer and Haywood in 1900 when he joined the union, and had become acquainted with George Pettibone when the storekeeper had sent two salesmen into Globeville to "peddle" wringers. They were arrested for not having a license and were told to turn to Malich who helped them out. At this particular time, there had been a strike at the Gobeville smelter for nearly two years. The smelter men's union was affiliated with the Western Federation of Miners, and venomous feelings were running high toward the employed non-union men. Malich was especially bitter and wanted action, so he proposed that the two mass murderers blow up a boarding house full of non-union help. They agreed, but planned the job on their own and neglected to obtain the approval of Haywood or Moyer, who were unaware of Malich's macabre proposal.

The pair located some Union Pacific powder magazines, and on a Saturday night before darkness fell, they pried the lock from the door and carted off six hundred pounds of No. 1 powder. Breaking into another magazine, they stole fifteen boxes of blasting caps. Now, they were in business again. A friend, Joe Mehalich, brought a team and wagon, loaded the explosives, and they were transported to the house where Adams and Orchard lived. They were then buried in the cellar.

When Haywood and Moyer learned what their prize henchmen were up to, they were quick to veto the proposed slaughter. There was enough work to do if they would only confine their work to targets already outlined. The matter was apparently settled, but a cellar full of explosives and an opportunity to experiment were more than the killers could bear. To make the most of their cache, they decided to develop a small hand bomb that could be hurled at an object and explode on impact. The plan had merit. Plaster-of-paris was molded around blasting caps and hardened. Nails had been tapped into each of the caps. When the crude bomb was thrown, the force of the impact drove the nails the rest of the way into the blasting caps and set them off.

The two dynamiters tested the hand bomb that weighed about three or four pounds. The base of the tree shattered as the blasting caps detonated. Adams was standing near the tree when the bomb exploded. Common nails and the metal of the blasting caps, deadly projectiles in flight, almost struck the surprised killer, warning them that the nail bombs had distinct disadvantages, as well as advantages.

Moving back into Denver proper, the killers were directed by Haywood and the always agressive Pettibone to make an assassination attempt on Justice Luther M. Goddard, Associate Justice of the Colorado State Supreme Court. Goddard had been appointed in January, 1905, by Peabody, and the Federation did not approve. The killers wore long overcoats that concealed sawed-off shotguns. The guns, broken into two main pieces, were hung under each arm by a shoulder brace. And off they went to exterminate another man.

The Goddard house, situated on the corner of Twenty-seventh Avenue and Humbolt Street, stood next to the sidewalk on one side and was set back a few feet on the other. The two killers casually strolled past the house and tried to determine if the jurist was home. Finally, using Mrs. Adams as a blind, the two men, accompanied by Joe Mehalich, caught sight of the judge at his door. Mrs. Adams saw him first and exclaimed, "There he is!" At that particular moment, Orchard caught sight of a young boy sitting on the curb and warned Mrs. Adams to shut-up. They walked away as if nothing had happened, and Goddard never realized that he had just missed being killed.

Joe Mehalich, a young Austrian smelter worker from Globeville, liked to visit the dangerous pair and play cards. Mehalich listened attentively night after night to the personal philosophy of the men who dealt in death and was impressed with their sage advice.

Toward the end of January, 1905, Steve Adams had gone on a drunk, and during the course of it, he was jailed for stealing a bicycle. At the time of his arrest, Adams was just sober enough to stash his Colt revolver in the patrol box. The weapon was later recovered, and ambitious newspapermen traced it to the Tritch Hardware Company, where a sales clerk identified the weapon as one of several that he had sold to a local merchant, George A. Pettibone. Adams was kept in detention for a week by the police and then released, since the law could not connect him with any particular crime, nor did they identify him with the notorious Steve Adams of Cripple Creek fame.

As the cold winter began to thaw, George Pettibone heard a rumor that James Peabody was going to be a guest of honor at a special dinner at the Elks Hall. He saw this as an opportunity to kill him, just as he was leaving the building and was relaxed, jovial, and buoyed up with the spirit of the occasion. Pettibone had been recommended for membership in the Elks by James J. Sullivan and had been initiated into the order just ten days after the destruction of the Independence depot. Pettibone had been "hooked" on joining the organization when he heard that Peabody, Wells, Goddard, and many prominent mine owners were members.

He stationed his two killers, armed with shotguns, on the street opposite the Elks Hall while he entered the lodge to check on the activities. A short time later he emerged, seething with anger. Peabody was not there and was not likely to be there, either. The dejected trio felt the need for a drink, a good strong drink!

In the meantime, union leadership was starting to think that someting could happen by having a drunk around with the knowledge that Adams possessed and by having him pulling cute tricks like stealing bicycles. Luck had been with them this time, but what about the next time? A man who drank like Steve Adams could get them all into trouble at any time. Harry Orchard, of all people, gave his friend a lecture on the evils of drink and what it would do to his position as a Federation assassin. It had little effect, as Adams soon splurged again and lost complete control of himself to such an extent that he had to be helped home.

The killer was coming unglued, not uncommon for men of the murder-for-hire profession. The tension and other conflicts were taking their tolls, and more and more liquor seemed to be the easy way to solve every problem. Both Haywood and Pettibone were now convinced that he "knew too much to be around drunk that way." One way or another, he had to go — it was just a question of when and how.

Circumstances became strained in the Adams' household, and Orchard, sensing that Adams was undependable and was becoming a stone around his neck, set out on his own the last of March. Still intent on "finishing" Justice Goddard, he moved to a room just two blocks from the Goddard residence. It was a good location, as he could see his victim board the street car every morning.

Adams seemed to sense that his former partner wanted to be rid of him, and he became quarrelsome. Orchard suggested that he team up with Joe Mehalich as it was a natural partnership since the two men and their wives were friendly. Adams, not satisfied with the suggestion, pressed the point. Grasping for a reason that would explain his coolness, Orchard faked a story that he was either going to Canon City or Colorado Springs to kill Peabody or Charley McNeil. "I'll go with you," Adams offered. Stuck for an answer, Orchard let the matter stand.

At the same time, Adams was having trouble securing money from the Federation. He was doled out small amounts, but Haywood had practically closed the purse strings on him. Mrs. Adams was allowed money for household expenses, but no money for drinking was available. The handwriting was on the wall, and the killer understood. He decided to pack up and leave while he still could.

Orchard went over to Globeville in early April to see his good friend off. He was leaving with Joe Mehalich. Orchard called Adams to one side in an attempt to make a little peace before they separated. They had not only killed men together, but they had been drinking buddies and good friends. Adams loosened up to his friend's genuine warmth, then exploded and blamed Orchard for his break with the Federation. He maintained that Orchard had run him down to Haywood and had destroyed the big man's confidence in him. Orchard shot back that the union leader knew about his getting drunk — everyone knew, for that matter. When the air cleared, Adams conceded that Orchard was right.

Steve Adams and his wife were broke. Orchard had little money with him, so he borrowed twenty dollars from Max Malich and gave that to Adams. He also instructed Malich to give them all the groceries they could use and then send the bill to Haywood. Orchard knew this was one bill that would be paid. It was a small price to pay to remove Adams from the country.

Adams gave him some relevant advice before leaving: "When they're through with you, they'll treat you the same way." Orchard considered the statement just sour grapes from a loser. But was it? In any event, the two killers would meet again in less than one year. It would be in another state and behind prison walls. And they would both be charged with murder — but that was in the future. Much more was going to happen before that, and Harry Orchard was going to have an important part in it.

CHAPTER 17.

Hogan was a good-natured Irishman.
. . . John L. Stearns

It was time for Harry Orchard alias Thomas Hogan to become an honest man — if only to keep up appearances. So the cunning George Pettibone arranged for an interview with John L. Stearns, manager of the Mutual Life Insurance Company. Thomas Hogan's impressive list of references included the merchant Pettibone, John C. Sullivan, president of the State Federation of Labor and an organizer for the Western Federation of Miners; Henry Cohen, an attorney for the United Mine Workers of America; another lawyer, James J. Sullivan; and Horace Hawkins, of the prodigious law firm of Richardson and Hawkins, also employed by the Western Federation of Miners.[52]

Stearns considered himself fortunate to obtain an agent with such good recommendations. His personal opinion was that "Hogan was a good-natured Irishman." He advanced his new agent fifty dollars to help get him started and was pleased when the man said he wanted to work the southern part of the state, near Canon City, which by an amazing coicidence happened to be the home of the former governor, James H. Peabody, who had now retired to private life.

Thomas Hogan, insurance agent, promptly arrived in Canon City on the Arkansas River, and like any other bona fide businessman checked into the local hotel. The next day, he went to work diligently selling insurance. At every opportunity he asked "innocent" questions concerning the whereabouts of the former governor. It was quite apparent that the salesman was enjoying his work. He liked the area so well that he moved from the hotel to a room in the home of an older lady, Mrs. Adams. The fact that his new abode was only a block from the Peabody home struck nobody as unusual. Orchard canvassed the town and learned that the former chief of state usually sat near a certain window at night to read, and that particular position happened to be located next to a vacant lot. He decided that a time bomb would be most effective if it were situated next to this window and set to explode with a delayed fuse. This way, there would be time for him to escape to a saloon and establish an alibi before the bomb went off. If everything went according to plan, Peabody's widow could soon use all the insurance her husband had purchased.

Orchard stayed almost a week in Canon City before he left for Denver to pick up the necessary components for his time bomb. After checking in at the Belmont Hotel, he told Pettibone his plans. Then he went to Globeville and picked up fifty pounds of dynamite and a box of blasting caps from Max Malich. The two men dug up the same explosives he and Adams had stolen when they had contemplated blowing up the boarding house. The killer put the explosives in a suitcase and transported it to Pettibone's store, stashing it in the basement. Harry Orchard wanted this bomb to be as professionally and technically perfect as he could make it. He was frustrated by his previous failures to kill Peabody and was determined to succeed this time, at all costs.

Across the street from Pettibone's new store at 1634 Court Place was the plumbing shop of Charles Roach.[53] Orchard told the plumber he wanted him to make a lead bucket about eight or nine inches in width and about fourteen inches high.

"What are you going to do with it?" asked Roach.

"I'm going to grow cactuses and bore holes in the sheet and have the cactus flowers grow up through them; its a great idea of mine," Orchard answered.

The whole idea was so strange that Roach remembered it for a long time, but he turned the job over to his assistant, Riley Harris. After it was finished, Orchard, with the assistance of Pettibone, took the lead cylinder and hammered the sides flat so it would lie without rolling. Then a wooden end was whittled and made to fit. Eventually, twenty-five pounds of dynamite would be crammed into the container. A hole would be cut in the top side of the bomb and a small amount of powder would be taken out; blasting caps would be inserted in this space, and a vial that contained acid was to be wired over the caps. An alarm clock, minus the alarm-bell, would be arranged so that when the clock's alarm went off, a fine wire would begin winding around the key, and in seconds it would pull the cork out of the vial so that the acid would run over the blasting caps and detonate the lethal mass.

While he was staying at the Belmont Hotel, the killer ran into a friend, William J. Vaughan. Vaughan was a former railroad man who knew the dynamiter under the name of Dempsey. When Orchard mentioned he was selling insurance and intended to make big money, his friend was interested. So Orchard not only suggested Vaughan join him in writing insurance, but also invited him to stay with him in Canon City. Surprisingly Vaughan was hired as an agent, and a few days later was on his way to join his friend Thomas Hogan in southern Colorado.

While he was rooming with the killer, Vaughan kept hearing a ticking noise, but couldn't locate it. Orchard was noncommittal about the noise, but when Vaughn left the room to go to the toilet and then returned, he found his partner fooling around with a clock. "He told me how he could use the clock to explode a dynamite bomb in a mine," explained Vaughan. Later, Vaughan remembered Orchard saying more. "In a joking way, Orchard told me he had a bomb in the grip and was going to put it under a man across the street." Vaughn took the statement with a grain of salt, but he later tried to move the piece of luggage and found it was extremely heavy. Another time, the killer pointed out the Peabody house and indicated he was going to use his bomb there. The salesman concluded his friend had a strange sense of humor, to say the least.

Toward the middle part of May, 1905, Orchard interrupted his lucrative sales career to make a trip to Denver to meet with Max Malich. This appointment had been prearranged on the murderer's last visit. Malich, still wrapped up with thoughts of arson, was considering burning a building he had insured, and he wanted Orchard's professional touch. When Orchard arrived, however, Malich was not quite ready to go through with it, so the murderer spent time with Pettibone catching up on the latest news.[54] The wily merchant informed his friend of the convention of the Western Federation of Miners that was to be held in Salt Lake City, Utah,[55] the latter part of the month. Pettibone was most concerned that "something be pulled off" before the convention opened, because too much money had been expended during the winter months, and the executive board had nothing to show for it.

Pettibone reflected on the Peabody problem and suggested Orchard give Justice Gabbert a try again, and since the killer could give no valid objection, he agreed. It is significant that neither Haywood nor Moyer were aware of this latest plot. Orchard was already acquainted with the judge's habits of walking back and forth to work. There was a vacant lot on one corner of his route, and across the lot was a beaten foot trail that the judge used as a short cut. It was a most appropriate spot for a killing; it was deserted, and a row of billboards shielded the lot from the street.

A dynamite bomb, crudely made by using a two quart molasses can as a bomb casing filled with dynamite and blasting caps, would be the instrument of death. It was buried in the earth just to the side of the trail and was camouflaged carefully with dirt and weeds. The original plan was to tie a woman's handbag to a windlass with acid bottles attached; when the handbag moved, the windlass would spill acid on the blasting caps and set off the detonation process.

When the bomb itself was set in place, the last essential detail was to attach the handbag to the windlass wire. By this time, Pettibone had become absorbed in the project and wanted to see for himself how well it worked. The next morning Orchard picked up the handbag which Pettibone had conveniently gotton from his home and stood on the edge of the vacant log, waiting for his victim. Pettibone was to signal when Gabbert was on his way — everything was ready. Gabbert came striding into the area as scheduled, but just at that moment another man cut in front of him.

The unexpected intrusion threw the killers' careful timing off. The men were walking too close together for Orchard to attach the handbag to the bomb wire, so all he could do was to let this moment pass and wait for another opportunity. Cursing, Orchard and Pettibone retired to the latter's store for whiskey and discussion. Pettibone was upset at missing the fireworks. It would have been almost comparable to the thrill he had had of shoving the powder down the wooden flume and blowing up the Frisco Mill in 1892. Reluctantly, he left for the convention.

Several more days passed without a presentable opportunity to kill the jurist. Then a new problem arose — the buried bomb became too sensitive to trust. The windlass might turn at the slightest touch. Because of his experience as a professional powderman, Orchard refused to work further with the buried explosives and certainly would not attempt to dig it up. The only logical way to deal with the situation was to blow it up with another bomb, a technique any well-trained demolition man would use.

So Harry Orchard, professional assassin, began his construction of a more elaborate bomb. It was a work of art, technically correct in every way. Noticing some eight-day clocks on the shelf in the basement of Pettibone's store, he began to form an idea. Taking a spring out of one clock, he tried to determine if it had enough force to break the glass vials he preferred to use for the acid in his bombs. He found, after several vials were broken, that the spring never failed to break the glass. So far, so good. Out of boards he found in the basement, he built a wooden box to house the explosive ingredients. Patiently, he fastened the clock spring along the underside of the cover by bending it back and then holding it in place with a piece of firm wire. This acted as a safety feature much in the same manner as the pin in a hand grenade. When the wire was removed, it released the clock spring which was then free to slam into the glass vials, breaking them, spilling sulphuric acid on the blasting caps, and exploding the dynamite. The wire was adjusted so that it slid out of the box with ease. It was a perfect set-up — worthy of praise from any revolutionary or guerilla leader.

The only remaining ingredient to be added was the explosive element, buried in the lot next to Pettibone's home. The dynamite used was that left over from constructing the Peabody bomb. It was a simple task to make a trip over to the "devil's" house to uncover the old door covering the hidden explosives and remove the necessary amount.

At night, borrowing Pettibone's bicycle, Orchard picked up his newest creation and peddled slowly to the death site. Parking the bicycle, he carefully dug a hole as close to the first bomb as he dared and placed the wooden box in it. Covering it, he used all the skills he possessed, realizing he would not be the first person to have an explosive device "go up" in his face. All that remained was to attach the handbag to the wire that jutted from the ground. Breathing hard, he wiped the sweat from his hands and face.

Unexpectedly, a stranger came across the path, advancing so suddenly that Orchard did not see him until he was standing right beside him. Reacting quickly, the killer feigned drunkness. The man spoke to him, but Orchard lay with his face to the ground, muttering and gagging, mumbling he'd be okay in a little while. Assured there was nothing wrong with the drunk, the stranger passed on.

Early the next morning the killer waited for a sign of Gabbert. And at last he appeared, heading for the vacant lot. Anticipating his victim's movement, Orchard got on the bicycle and headed for the bomb site, stopping only long enough to connect the handbag to the wire protruding from the buried bomb. Carefully he made the tie and hurried away as fast as he could. A block away he stopped, expecting to hear the report of the explosion. But he heard nothing!

At the rate of speed Gabbert had been traveling, he should have reached the booby-trap. Something was wrong! Didn't he see the handbag? Why hadn't it exploded? The killer knew he should go back and see for himself, but instead he peddled toward the downtown area. What had gone wrong?

About an hour later, the entombed death box exploded, and its power reduced a living human to a mass of shredded bone and flesh. And what actually happened was indeed more bizarre than fiction. For just as the intended victim came to the edge of the path through the vacant lot, something told him not to go that way. Instead of pursuing his traditional route, the justice returned to the sidewalk and walked straight to Colfax Avenue and to the state house. Arriving at his chamber, Gabbert met Justice Goddard and told him about the unusual experience he had had, adding that he probably was very silly and superstitious.

What actually took place that May morning of the 24th was that Merritt B. Walley, an unemployed mining engineer from the Cripple Creek district, was in town, job hunting. Crossing the vacant lot, he evidently spied the handbag on the ground, bent down to examine it, and as a local newspaper reported the incident, the resulting explosion "blew his tired soul into eternity."

The force of the explosion from the two ten-pound bombs threw his body thirty feet in the air and flung his vital parts in all directions. So great was the burst that the only recognizable portion of his body was the trunk. Volunteers gathered pieces of the body in a dishpan. Feet and shoes were found a block and a half away; a finger lay on the lawn of a dwelling a block from the blast area. It was a fast, but messy way to go.

The most prevalent explanation of what took place, but one that the family and close friends denied, was that the victim had committed suicide because of unemployment and his wife's illness. Five minutes after his death, a telephone message, notifying him of a possible job in the city, came to his house. Ironically, Walley had been arrested on July 24, 1904, for complicity in a plot to blow up the Portland mine where he was employed as a mining engineer, and had been confined in the military bullpen for a number of days. When he had refused to take out a card in the Mine Owners Association, he was ordered out of the district.

Orchard had again destroyed the wrong target. He had failed to eliminate Bradley, whose recovery was now assured, and now he had killed an unknown man who supported the very organization that had murdered him. Since his return from California in December, Orchard realized he had produced nothing constructive for his employers. Even now the unused bomb earmarked for Peabody was sitting under Mrs. Adams' table in his stored suitcase. He had to redeem himself; his reputation as a professional was at stake. He had to make it big, next time. The killer now set his bomb sights again on Justice Luther M. Goddard. Only an act of God could save the justice now, Orchard swore.

CHAPTER 18

He went armed like a Cuban Major General.
. . . William E. Borah

This time, he told himself, he would not fail to get his man. And so Harry Orchard set to work on an instrument of death. The bomb casing was a simple wooden-pine box, containing the usual ingredients — forty sticks of dynamite on which were placed one hundred blasting caps. Packed solidly around these caps was cotton saturated with potassium chlorate and sugar.

On top of the wooden box and tilted downward was a small glass vial filled with sulphuric acid. This vial had a rubber cork through which a strong pin was run, the head of the pin flush against the bottom of the cork. The other end of the pin was bent to form a hook and was attached to a cord, the end of which was to be fastened to a screw-eye embedded in Judge Goddard's front gate. The booby-trap would be activated when the gate was opened. The pressure of the cord would pull out the cork in the vial of acid and allow the contents to run through a small funnel-like opening in the top of the wooden box, where they would make contact with the saturated cotton and burst into flames to explode the blasting caps and powder.

Proud of his craftsmanship, Orchard went to Globeville to tell Max Malich about the new bomb and even showed it to him. The next day, shortly before daybreak, an Austrian friend of Malich drove the killer and his deadly cargo into Denver and to the residence of Associate Justice Goddard. The dynamiter jumped out of the rig and with a shovel started to dig a hole close to the gate post. Silently but quickly he peeled back the grass and removed the dirt. Lowering the wooden box into the hole, he carefully tapped the loose dirt around it, then replaced the grass. He made sure there was no tell-tale evidence. The green fishline cord extending from the buried bomb was not tied now, but it was left lying on the ground covered with cut grass. This was done as a precautionary measure to insure that nobody but the victim would open the gate once the fishline was attached. Orchard returned to his driver and rig, and they went back to Globeville to have breakfast with Malich. He had worked up a good appetite.

The second part of the plan was to take a streetcar back to Denver and finish connecting the cord of the bomb to the gate. A little past eight o'clock, Orchard knelt in front of Goddard's gate. He had dropped the newspaper he carried to make his action appear accidental, and while he was stooping to pick it up, he tied the cord to the screw-eye set in the wood of the gate. Nonchalantly he walked away, found a streetcar to the downtown area, and waited for the anticipated sound of the explosion. He heard nothing. Checking his watch, he reasoned that if Goddard were on schedule, he should have passed through the front gate by now. Minutes turned into hours, and still nothing happened. It was evident that something had gone wrong. Had he been spotted when he attached the bomb cork to the gate and had the judge been warned that something was wrong?

By sheer luck, the deadly mass of explosives never did detonate. The powerful acid had eaten the head off the pin that passed through the cork, and when the judge had opened the gate, the pin pulled out and left the cork in place. The head of the pin had corroded in the cork and sealed up the hole, so there was no leakage. Later the grass covering the bomb died and gave the spot a most unhealthy look, so that the instrument itself remained undetected for nine months.

When the Goddard bomb failed to detonate, Orchard felt sure he had run "clean out of luck." Everything seemed to be against him. Sullen and morose, he returned to the San Luis Valley where Vaughan was still writing insurance policies. Orchard went to work at once selling insurance with the zeal of a high-powered agent, as if he had to prove himself after the Goddard failure. And he did prove to be a fine salesman. During a two-week period in June, he made the kind of money that insurance agents like to brag about, even though killing was easier and much more exciting.

When the union leaders returned from the Federation convention in Salt Lake City the latter part of June, none other than Harry Orchard was on hand to greet them. The important outcome of the convention had been the re-election of Haywood and Moyer. When Orchard gave them a rundown on his most recent action, they advised the frustrated assassin to cool the murder routine for the time being and lie low. Since the union leaders were preparing for a trip to a Chicago convention to help finish organizing the Industrial Workers of the World, it might be better to keep Orchard down in the southern part of the state until he was needed for another job.

In June, from Monte Vista, Orchard sent Stearns, manager of the Mutual Life Insurance Company, the fifty dollars he had borrowed in April when he had started to work. Orchard gave as his reason for quitting that he could not make a go of it. Then, during part of June and July, Orchard, Vaughan, and another man named McGiven worked for H.C. Valentine, an agent for the St. Paul Mutual Hail and Cyclone Insurance Company. Valentine was more than satisfied with the murderer's efforts as an insurance writer, and Orchard earned more than five or six hundred dollars during his brief tenure with the company.

Returning to Denver after his selling sojourn in the southern part of the state, Orchard wasted no time in getting together with Pettibone to discuss the failure of the Goddard bomb. The pair concluded by agreeing that the "killing game" would progress more rapidly and with better results if they had a readily available means of transportation. They took the problem to Haywood, who said he would get them a wagon and team.

Pettibone had brought up the matter of eliminating the former Adjutant General of the State Militia, Sherman Bell. The over-aggressive and often abusive action of Bell in the mining strikes could never be forgiven. Orchard and Pettibone conducted a thorough survey of the intended victim's home on Thirteenth Avenue, near Congress Park, making their forays in the new buggy that Haywood had let them purchase for the job.

Despite his preparations, a good opportunity to kill Sherman Bell never presented itself. Orchard had decided to use a shotgun for the act, but to use this type of weapon, closeness was essential. Bell was seen on the porch on several occasions, but it was dark where he chose to sit and comparatively light where his assassins would have to stand in order to shoot him. Several times Orchard tried to approach the house, but dogs began to bark and cause an uproar. At these times, the ever-reliable Pettibone would be waiting in the buggy down the street for the quick getaway. He was never one to expose himself to unnecessary risks.

It was often speculated that in time Sherman Bell could have been killed, but any further attempt was vetoed by Moyer himself, and he called off the killers. The union president reasoned that if anything happened to the old soldier, he would "sure as hell" be in jail for it, which was probably right. Moyer complained his health just could not endure another imprisonment like the Telluride affair had brought. At this point, Moyer opposed any detrimental involvement that could be blamed on the Federation.

One man in particular, however, worried Moyer. He knew too much and could blow the whistle on them all, anytime. Moyer was losing sleep because of his obsession. The man who kept the leader of the most active and violent union in the United States awake at night was none other than Harry Orchard's old comrade, John Neville. They had to get that son-of-a-bitch, for the good of the Federation. Oddly enough, Moyer was the only person really concerned about Neville.

Meanwhile, Ida Toney Orchard had been badgering Haywood about her husband's whereabouts. Again Haywood contacted the mass slayer and in no uncertain terms instructed him to take care of his feisty wife before she drove him crazy and before she accidently spotted her husband on the streets. Orchard was amused to see the big man so worked up over a little woman like his wife. If she were not so much of a drag on Harry's nerves, he would have found the situation humorous. A temporary solution presented itself when the dynamiter visited Marion Moore, a member of the executive board of the Federation. Moore was leaving for Alaska to organize a union at Nome, and Orchard saw a new way to gracefully evade his wife once more. He asked Moore to mail a letter for him from Nome.

Orchard addressed the letter from Cape Nome, Alaska. He started out with the salutation, "Well, Dolly," and then followed with, "I will not call you my wife as you are ashamed of the name." If Ida did not perceive the implication as to her status, she never would. The killer went on to announce that he was in Alaska to make a fortune, but he gave no hint of the future. He handed the letter over to Moore, who noticed that it was addressed to Mrs. Harry Orchard. Interested, he asked why, and Orchard told him he wanted to get rid of "that woman." When the Federation organizer arrived in Nome on August 12, 1905, he mailed the letter.

At the same time, the relationship between Moyer and Haywood was becoming seriously strained. They were drifting apart, and it was noticeable to everyone. Moyer was becoming conservative in his labor dealings; evidently, he had had enough of jails and bull-pens. Big Bill, on the other hand, relished even more the physical confrontation with opponents and running the big strikes. The violence of extremely emotional men had a certain appeal to him. The break between the two men seemed to have started with Moyer's imprisonment at Telluride, when Haywood thought Moyer was going soft and was cooperating too much with the hated military.[56] No doubt the publicity Haywood drew and his rugged appeal to the eastern people annoyed Moyer. He didn't want Haywood stealing the scene.

August, 1905, was a strange month for the group. Most of the time Moyer was in California with his wife having had her send him a telegram saying she was sick and asking him to come, Haywood was drinking as if there were no tomorrow, disgusted with Moyer's deceit which he had discovered, and Pettibone's wife was in the east visiting. Orchard spent most of that month living with Pettibone. This gave the two men plenty of time to concentrate on murderous activities and even to talk about retirement. It was a great month for Harry Orchard.

During the time Moyer was in California, Haywood went to pieces, but he pulled himself together after Moyer returned. The two men managed to work together fairly well on Federation business, but had little in common after working hours. Regardless of how the two union leaders felt about each other, Pettibone continued his friendship with both and at times even cauterized their differences. He apparently did not allow his wife's brief affair with Haywood to interfere with Federation activities. Maybe Pettibone's need to be a member of the "Inner Circle" was more important to him than were his wife's indiscreet actions. He was also well aware that without their friendship he would be just another Denver businessman.

During the last week in August, Moyer held an informal

conference in the Federation offices with Haywood, Pettibone, and their "hit or miss" assassin, Harry Orchard. Moyer stressed his edict of no further violence in the "Mile High City," but Pettibone objected. He felt that the present was the most opportune time to repay past dues. Haywood was willing to go along with the merchant and take his chances, but Moyer was adamant on the point. No rough stuff in Denver, he insisted.

The Federation president proposed as an alternative an effort to shut up that cheap blackmailer, John Neville! Neville was dangerous and likely to be an unstable man if he ran into trouble. One solution was to kill him! Moyer was the only one in the group who really felt Neville was important. In fact, Haywood put forth his candidate to be eliminated — Frank Steunenberg of Idaho, the man who had been governor during the Coeur d'Alene riots. Haywood "grieved" for Steunenberg's death with a passion, and on various occasions he had even sent Steve Adams, Art Baston, and Ed Minster, all Federation henchmen from the Cripple Creek district, and a hatchetman from the Coeur d'Alene district by the name of McCarthy, to kill him, but they had all failed.[57]

Moyer agreed to the Steunenberg kill and suggested that a special macabre touch be added. Once the former governor was slain and the crime was known to the world, they could write letters to Peabody, Gabbert, Moffatt, Bell, and Hearne, and other enemies of the Federation as a reminder that they had not been forgotten and that it was just a matter of time before their deaths would also be announced. Haywood pushed the scheme even farther by recommending that the letters be mailed from Patterson, New Jersey, to give the impression that foreign anarchists had sent them!

Pettibone reminded them that the former governor lived in Caldwell, Idaho, a small rural community. It might be difficult for a stranger to move around, commit murder, and then escape successfully. Haywood nodded his head in agreement, but added that the information they had on Steunenberg was that he was engaged in the sheep business, and so he moved around visiting isolated sheep camps. That would provide them with an excellent opportunity to set up an ambush to kill. Chances were also good that the victim thought the Coeur d'Alene riots were a thing of the past and would not even expect an attack on his life.

Orchard selected the former governor as his first target, hoping to put off the Neville affair as long as possible, even though Moyer still wanted Neville eliminated. At any rate, the conference ended with the murderer collecting two hundred and forty dollars expense money to get him on his way. But before he left, Pettibone disregarded Moyer's caution and talked Orchard into making one last attempt on Sherman Bell's life. The killer did not want to oppose Moyer, but at the same time he wanted to appease the merchant. So Orchard only pretended the motions of attempting to shoot him. Pettibone was downcast about his not making the kill, but he still had no desire to become further involved himself.

This last attempt at Sherman Bell clearly shows that the violence generated by the members of the Western Federation of Miners was not always the close, well-disciplined power people assumed it was. As with all organization, the most ruthless acts are not always instigated by the top leaders, but by fringe elements. And in this case, the attempt against Bell was done in direct opposition to the desires of the president.

The next day, following the unsuccessful attempt at murder, the dynamiter packed his bags and purchased a round trip ticket to Portland, Oregon. Included in his itinerary was a side excursion to Seattle and the Puget Sound area. A former Cripple Creek miner named Arthur Parker had settled in the region, liked it, and had recommended the area. Pettibone and Orchard had given serious thought to buying a place on the Canadian border where they could settle down and relax in peace. They could also do a little border smuggling and harbor wanted Federation men on the side. The pair thought Washington offered more advantages for this activity than did southern California. Orchard had approached Haywood with the idea. He liked it and even promised the necessary financial support.

Orchard dug up the buried Peabody bomb which he had buried in a manure pile at the barn where the horse and the rig were stabled and was taking it with him. An explosive device built for one governor and used on another governor was too good to pass up, he felt. Perhaps this time it would bring him better luck. In any event, before he left, the point was made rather strongly to Orchard that if he were to run into any kind of trouble, he was not to wire, write, or attempt to contact any of them or the Federation. They would see to it that he was provided with a lawyer at the proper time.

Harry Orchard realized the Idaho adventure would be a big one and a difficult one, so he prepared accordingly. Thinking he might have use for a "Pettibone's dope," he purchased a seventy-five cent bottle of stick phosphorus and a twenty-five cent bottle of bisulphide. All he had to do was add alcohol, benzine, and spirits of turpentine. The killer did not want to get all the ingredients together because a druggist might question the use. He also planned to take along with the Peabody bomb, a pump shotgun, some brass knuckles, and a Colt automatic. This assortment of hardware later caused Borah to comment, "He went armed like a Cuban major general."

By the time Orchard got everything ready, he had enough to fill a trunk, a suitcase, and a valise or traveling bag. He put the bomb in the trunk and then packed the rest of his arsenal, adding such items as bedding, clothes, ammunition, medicine, plaster of paris, a ladies handbag, fishing reel, crimpers, a pack of business cards marked Thomas Hogan, Silverton, Colorado, agent, Mutual Life Insurance Company, various other articles, and strangest of all, pictures of his baby, Olive, that he had brought with him from Canada. No wonder Pettibone had to help him to the station. They were both relieved when the trunk was checked, loaded, and secured for travel.

Finally, sometime between the twenty-fifth and thirtieth day of August 1905, the union assassin climbed aboard the Rio Grande railroad car which would take him to Idaho where he was to commit what he claimed was "my crowning crime."

CHAPTER 19.

Dynamite by God!
. . . Jack Simpkins

Bent on murder, Harry Orchard headed relentlessly for Nampa, Idaho, a small community west of Boise, the state capital. Nampa was situated about nine miles from the target city of Caldwell, a perfect place for gathering background information on Caldwell and his victim before he actually finalized his plans. He checked into the Commercial Hotel and set out to learn all he could. After a couple of days in Nampa, the killer moved on to Caldwell, a quiet, rural community of some three thousand people, located near the Oregon border and just twenty-nine miles from Boise. It was a most unlikely place for murder.

Registering at the Pacific Hotel on September 2, he was given room seven under the name of Hogan. He told the manager, Stephen Dempsey, that his stay would be but a few days in duration and that he was planning to buy some sheep. Whom could he contact that might help him? Orchard was given a list of some of the successful sheepmen in the area; the list included the name of Frank Steunenberg. When he asked further about the former governor, he was told the man was not around Caldwell much, but he was generally in Boise or in Mountain Home to buy and sell sheep.

Even though his victim was not in the area, it gave the killer an opportunity to become more familiar with the town and to locate the Steunenberg family residence. Satisfied he had learned all he needed, Orchard took the train to Boise where he had been told the former governor was conducting some business transactions. Arriving in the city, he checked into the Capitol Hotel. As he signed the register, the assassin made a quick search through the guests' signatures to see if Steunenberg's name was listed; it was not.

The following morning, determined to locate his victim, Orchard strolled over to the city's leading hotel, a splendid structure named the Idan-ha. Lounging casually around the check-in counter, he saw an opportunity to look at the register while the clerk's back was turned. Running his finger along the guest list, he saw the signature — his target was registered here.

Registering at the Idan-ha, Orchard managed to get a room on the same floor as his intended victim. At noon when the chambermaids were off the floor for lunch, the killer tried a skeleton key in Steunenberg's door, and it worked. Satisfied, he returned to the hotel's lobby to consider his problem and his next move. Again, by chance, he struck up a conversation with a sheep-buyer from Wyoming. He told the man that he was in the same business, and they hit it off. The salesman took him across the street to meet a Mr. Johnson, who also happened to be in the sheep business. Quite naturally, the talk dealt with animal operations in the general area. Orchard's new found friend interrupted the conversation to say he had formerly bought some sheep from a man by the name of Steunenberg. Yes, conceded Johnson, the former governor was very active in the industry. "By the way, that's him over there in front of the Idan-ha now," said Johnson as he pointed the man out. Orchard got his first look at the man he was going to kill. He looked like any other mortal, not a fiercesome or hateful object at all, Orchard casually observed. Orchard and his friend returned to the confines of the hotel, and while the Wyoming salesman talked, the killer began thinking of the ways he could dispose of the victim. Perhaps the Peabody bomb packed in a trunk and stored in the Nampa train depot could be placed in Steunenberg's room and effectively detonated by an alarm clock.

The dynamiter was well aware that the explosive charge would tear the hotel apart and possibly result in other persons' dying needlessly, but that possibility was of little concern in the planning of the assassination — other innocent people had died before in his operations. What was of concern to the professional killer was the threat of capture. Fear inevitably was his predominant emotion as he closed in on his victim. The more he thought about it, the more his confidence began to crumble. He realized the time bomb would give him a certain time period to make his escape. Possibly he could even be half way to Portland before it exploded. The more he thought about his being discovered, the more he wanted to be away from Boise and Steunenberg, as soon as possible.

Orchard's decision was made on the way to Nampa to get the bomb. He would make the trip to Seattle and hunt around for the ranch he and Pettibone had talked about. Suddenly, the little ranch became important; the thought of it consumed him. He must go! Boarding the Portland-bound train that very night, he gave the former governor a four-month extension on life.

It was now the middle part of September. The cold, wet weather of the Seattle area dampened his enthusiasm; it depressed him further. He did have, however, a commitment to Pettibone to try to locate an ideal, secluded ranch. He found a friend of Pettibone's by the name of William Barrett, whose mailing address was a saloon on Pike Street. Over drinks, Orchard told him of his trying to locate a small place near the Canadian border on land that was heavily timbered and had a good water front. Barrett said he could help by introducing Orchard to some real estate people he knew. A week passed as the killer was carted all over the countryside to look at real estate, but the depressing effect of the climate urged him to go on. He asked Barrett to send a map of the Puget Sound region to Pettibone to let him make the decision.

Orchard then left for Spokane, then continued farther east to the mining region of the Coeur d'Alenes. Orchard had arrived in the district years ago as an arsonist and a wife deserter, only to flee the mountain region after participating in a murderous mob action. Now, he was returning to his old stamping-ground as a professional assassin, a mass murderer.

As the train wheels hit the metal of the rails, Harry Orchard considered adding a new slant to his criminal career — kidnapping. It would be a hell of an easier way to make money than killing, and less nerve wracking. David Coates, a former lieutenant-governor of Colorado and an official in the Federation, had suggested the kidnapping plot to Orchard. Coates was preparing to leave Colorado for Wallace, Idaho, to take over the publishing of a local newspaper, *The Wallace Tribune*. He had been introduced to Orchard by Pettibone, about the time that Haywood and Moyer came back from the Chicago convention. Orchard was using the name of Thomas Hogan. When the killer learned where Coates was moving, he became interested. As Coates recalled the incident:

> Orchard walked down the street with me, saying he had once lived near the town of Wallace, in the Coeur d'Alenes. He said he once had an interest in the Hercules mine and that Paulsen was one of the owners . He said he was known in the Coeur d'Alenes as Harry Orchard, and asked me to give Paulsen his regards and to tell him he was coming soon.

It was during this talk, apparently, that the former politician proposed the kidnapping scheme. From later actions, it appears that Coates had been merely role-playing and wanted no part of it. In fact, when the real showdown came over the proposal, Coates was scared out of his wits even to possess any knowledge of it.

For Harry Orchard, the more he considered the possibility of kidnapping his good friend's children, the more appealing the idea became, particularly when he was planning to ask sixty thousand in ransom. Since his nerve had begun to crack in Boise, the ideas sounded even more attractive. Kidnapping was certainly better than killing Steunenberg in the Idan-ha Hotel. The fact that Paulsen was one of his best friends and that he truly held the man in genuine affection was a minor detail. Perhaps in his warped mind he considered the sixty thousand rightfully his. Orchard brooded over the fact that his former partners in the Hercules mine were wealthy individuals, while he lived from handout to handout. The circumstances under which he had sold his shares in the mine didn't bother the killer; it was just that his former associates in the Hercules were wealthy, and he was not.

Orchard's first stop in the Coeur d'Alenes was in Wardner, where years ago he had reduced the mighty Bunker Hill and Sullivan concentrator to a pile of rubbish. It was also the home of Jack Simpkins, presently a member of the executive board of the Federation. More important, he was one of those powerful individuals who made up the Inner Circle, and in this capacity Simpkins had to oversee the mining districts of Oregon, Idaho, and Washington. Over some drinks of raw whiskey, the killer laid on the line to Simpkins his reason for being in Idaho. Did Simpkins want a piece of the action? You bet he did! He wanted to be in on both the kidnapping and the Steunenberg killing. So together, they went to Wallace to visit Coates.[58]

While Simpkins walked into the house to get Coates, Orchard remained outside. When Simpkins mentioned his reason for the visit, Coates managed to stammer that he was only "joshing" about the whole affair. The three met later at Coates' office, and he made it even plainer that he wanted no part of such an insane plan. He did agree, however, to find out how much money Paulsen kept on deposit in the Commercial Bank. Paulsen was also Coates' neighbor; they lived across the street from each other and very likely the Paulsen children played in his yard.

A few days later Orchard saw Coates again and mentioned that he had visited Paulsen's home, had eaten dinner there, and had played with the children. Angry, Coates began to comprehend the enormity of such a crime and was determined that his own connection must be severed. "Orchard, you're a fool, and I would be a bigger fool to go in with you in such a crime." Defending himself further and fearing that he might be drawn in deeper with the intended kidnapping, Coates threatened the killer. "If you try to do anything like that around here, I'll denounce you."

"Oh, you needn't get so excited about it," Orchard assured him as he walked away.

With Coates out of the picture, they now had to replace him with another man. The plan would involve three persons to handle the kidnapping job properly. The third man Orchard and Simpkins decided upon was the rabid union-organizer, Vincent St. John, the terror of Telluride. St. John had been living underground since he ran away from Colorado and a murder warrant. He was now living in Burke and was working as a common miner in the Hercules, under the name of John W. Vincent. Unable to leave union activities, he also served as president of the Burke union. Since St. John and his wife had no children of their own, Orchard decided they could care for the Paulsen children while they were being held for ransom. A message was sent to Burke, and the union organizer met with the conspirators. He immediately rejected the proposal. However, St. John did recommend a hoodlum from Leadville, Colorado, by the name of Cunningham for the job. Cunningham was supposedly working at the Tamarack mine, but he would probably quit and join them in Wallace.

Orchard and Simpkins tramped around Placer Creek outside of Wallace and located a place in a heavy thicket where Cunningham could hide the children. Simpkins' assignment was to remain in sight of the spot where the ransom money was to be placed. When he was satisfied that the money was there, he would signal Cunningham to release the children. This exchange would take place in the late afternoon or dusk so the kidnappers could see what was taking place, and if necessary, could escape under the cover of darkness.

The plan called for the kidnappers to enter the family house at night when the Paulsens were away and seize the children from the nurse. A letter, telling them where to leave the money, would be left behind. Orchard was to be the lookout and check for evidence of a posse. When it was over, Orchard and Simpkins would head for the latter's timber claim on Marble Creek and "hide out" there, while Cunningham would return to Burke and play it from there.

They finally located the elusive Cunningham at the Tiger and Poorman mine. He was not the most stable man, but he did agree to become a part of the wild scheme, and at this point they had little choice. With him, their plans began materializing, but at the last minute mother nature became their opponent. Snow fell in the mountains, and rain poured in the valley. Discouraged, they admitted the whole scheme was too risky. They would have to keep the Paulsen children in damp, cold weather. Not wishing them bodily harm, they reluctantly gave up the escapade.

At this point, Orchard's criminal career received another setback. He was not only dragging mentally, but worse yet, he was desperate for money. Orchard's envy of his old mining partners made him bitter because he had had to leave the mining district. In his attempt to wrestle with the truth and perhaps build up false courage for the forthcoming murder mission, he rattled off irrational statements that he would "get even" with that bastard, Steunenberg. He considered Steunenberg directly responsible for all the trouble in the Coeur d'Alenes, since he had sided with the mine owners against the Federation.

Despondent, the killer trudged down to a combination barber

shop and loan office of John Kammerer in Wallace and pawned several items. On September 22, Orchard borrowed money on a revolver, another time on a watch, then on a watch chain, and finally on a pair of field glasses. The money Orchard received went fast on drinking, gambling, and what the dynamiter discretely termed "sporting." Attempts to get money from the Federation headquarters had come to nothing. Action first, pay later! They were reluctant to buy the "tourist bit" at company expense, not even for men who murdered for them.

Sheriff C.H. Williams of Shoshone county arrested Harry Orchard on the streets of Mullan as a suspicious-looking character. The murderer was released when he proved to the satisfaction of Sheriff Williams that he was gainfully employed by working for a man and running a wheel of fortune in a local saloon. Since Orchard had a natural aversion to work, taking bets and spinning the wheel of fortune must not have been too demanding.

Trying to pick up a little spending money, Orchard met a jewelry peddler by the name of Scott, who was interested in purchasing some Moonlight mining stock. The killer bought five shares of Moonlight for him and earned a commission of twenty-five dollars. Scott was a traveling jewelry salesman who made regular stops in Wallace. He sold to individuals only and carried valuable stock with him, usually amounting to well over a thousand dollars. He became acquainted with Orchard who then introduced him to Jack Simpkins. As the men became friendly, Scott was taken around and introduced to local business people, and subsequently made a great many sales.

As a reward for his assistance, the salesman lent Orchard money on several instances and even presented Orchard and Simpkins with watches. But it seemed as if the kindness of Scott seemed only to whet the thirst for his money. Orchard and Simpkins tried to steal Scott's sample trunk from the O.R. & N. Depot. They erred in the selection of baggage and ended up with a useless collection of shoes. Frustrated, Orchard went to Burke and stole forty dollars from the cash register of a friend who ran a saloon. Harry Orchard was fortunate he always seemed to find a friend to help him out of a tight spot.

During the planning stage of the kidnap plot at the Bi-Metallic

Hotel, Cunningham rented a room at the establishment to be near the other partners in the crime. Toward the end when everything was falling to pieces and they decided to scrub the whole affair, it was evident that nobody was going to get anything out of it. Angered, Cunningham broke into Orchard's room and cut open his grip, removing his revolver and other articles. Orchard caught him in the act and made him return everything. Trying to pacify Cunningham, the murderer gave him the Peabody bomb which Simpkins had ridiculed as being too clumsy and likely to attract attention.

The bomb had been made large at the suggestion of Haywood, Moyer, and Pettibone, who maintained that if the Bradley bomb had been larger, it would have killed him. Orchard had also attached two dozen one and one-half inch bolts to the outside of the bomb casing so that they would be thrown everywhere when it exploded, increasing the effectiveness in the killing zone. Cunningham had use for the bomb, as he was considering blowing up the Tiger and Poorman boarding house because scabs were living there. This was the perfect way for Orchard to get rid of the Peabody bomb and have it used for something constructive at the same time; thus Cunningham became the proud possessor of a big, lethal mass of explosives enclosed in metal.

The only possibility of his receiving cash on the spot was from the very man Orchard had plotted a crime against, August Paulsen.[59] The killer was brazen enough to put the touch on Paulsen by telling him a cock-and-bull story about needing financial help to get to California so he could engage in selling mining stock, which that good citizen, Dave Coates, was supposedly going to furnish. Paulsen did not question the soundness of the proposal. He gave his former partner the requested three hundred dollars.

Orchard returned to the Wallace loan shop and redeemed everything he had pawned, paying for them out of a roll of twenties which he flashed for the benefit of the store proprietor. After this, he played around for a few weeks, and then he and Simpkins left Wallace in the middle of October for Spokane. After staying there several days, they decided against wasting any more time and struck out for Simpkins' timber claim on the upper St. Joe River near Mica Meadows in northern Idaho to enjoy some relaxing hunting and fishing. It was a leisurely rest—drinking and taking life easy. The high point of it all was described by Orchard:

> During the time we were over there Jack showed me where he and Adams and the other killed Boule, the year before, and his horse and dog, and where the other claim jumper that was with him ran as they were shooting at him.

They examined the burial spot. Adams had already told Orchard of his part in the killing, and now Simpkins rehashed the murder for him. It had been successful, and the victims' bodies were safely buried. This was a different kind of disposal than Orchard was used to: he always left the bodies of his victims littering the scene.

The latter part of October found the two murderers in Spokane, staying at Simpkins' house on the north side. Simpkins was enthusiastic about eliminating Steunenberg and wanted a part in the proposed slaying. The more they talked, the more excited Simpkins became, and the more confidence Orchard felt. He needed someone to bolster his damaged ego, someone to egg him on and to supply courage. Actually, Simpkins could make an official trip to the area in his capacity as a member of the executive board of the Western Federation of Miners, and if he called on union leaders, he could charge the Federation with his time and expenses.

After leaving Spokane, the killers went straight to Caldwell and checked into the Pacific Hotel. It was the third of November. Registering as Hogan and Simmons, they requested a room together. They got the best in the hotel, No. 1. After searching for the former governor and learning nothing concerning his whereabouts, the two men went to Nampa and telephoned his residence. Orchard learned the victim was in town, but was not home at that time. Elated, the murderers immediately returned to Caldwell.

Hogan and Simmons, as they called themselves, went to work. In the reflection of a bright moonlight night, the two men paid a visit to the Steunenberg house, and using a pair of French opera glasses, they scanned the house for some sign of the victim's presence. Some of the window blinds were up, but the assassins could not determine for certain that Steunenberg was actually in the house. Returning to the hotel, they finally agreed upon a plan for perpetrating a perfect murder.

The next day they constructed a bomb to be used as a booby-trap. The casing was a wooden box built around the dynamite Simpkins carried with him and containing the usual Orchard ingredients — sulphuric acid, a windlass, and plenty of blasting caps. They decided to eliminate their victim under the cover of darkness at some point along the path leading to his house. A cord or wire could be stretched across the path so that when the victim ran into it, the explosive device would be triggered. Since the Steunenberg residence was the only home on this particular street, it was unlikely that anyone else but the former governor would be using the route late at night.

Steunenberg was found Sunday afternoon in the office of the Saratoga Hotel. The murderous pair watched his movements until it was dark outside, and then assuring themselves that he would be returning home soon, they took the box of instant death and headed down the path leading to the Steunenberg house. A short distance from the structure they halted. The bomb was laid on the ground while Orchard scooped dirt out of the hole close to the path itself. When he was certain it was deep enough, he gingerly placed the bomb in the cleared space, and heaped dirt around it. Weeds were scattered over the top of it as camouflage. A thin wire, attached to the bomb's windlass containing the sulphuric acid, was strung across the path and fastened securely on the opposite side to a stake. The booby-trap was ready.

The dynamiters dusted themselves off and quickly returned to the Pacific Hotel to establish an alibi. Nervously they waited for the anticipated explosion, but it never came. One hour, then another went by, and still no explosion report. Unable to endure the tension any longer, they went over to the Saratoga Hotel to see if Steunenberg had left; he had. Leaving the Saratoga, they traced the route back to the bomb site and saw that someone had passed by and broken the wire. Examining the explosive device, they saw that the force used to break the trip wire had turned the windlass with the acid in it so fast that not a particle of the chemical had spilt on the blasting caps. A small miracle had taken place.

Checking the bomb further, they saw that the bottle of acid was nearly back to its original upright position, but it was still on a slight angle so that the acid was ready to spill out, drop by drop. A moment of uncertainty followed — should they leave the bomb and let it explode or try to prevent it? Finally, Orchard put his finger over the mouth of the glass vial and then dug the explosive device out of the ground with his free hand. Simpkins bent down to help, and together they lifted the bomb out and carried it to a place near the railroad tracks. There they buried it again, covering the spot with weeds and other debris. Keyed up over the night's activities, they returned to their room where nervous exhaustion overcame them and they slept soundly.

For the next few days the dynamiters hunted the Federation's target, but they could not make connections with him. Simpkins began to feel the pressure and became afraid. He also felt he was too well known to stay around any longer without arousing suspicion. The Federation official decided the time was right to take care of some union business in Silver City, but not before he urged Orchard to stay on the project until the assigned mission was completed. Assured that important people still had faith in his abilities and that he was not forgotten, the killer saw Simpkins off. He fully realized that his status as a professional assassin would be enhanced if he succeeded in destroying the former governor. He could also demand more money, so Harry Orchard stayed behind to take another human life, certain that this time the bomb would not fail to go off at the right time.

CHAPTER 20.

Orchard was about as determined a countenance as I have ever seen on a human being, with the most cold, cruel eyes I remember having seen . . .
. . . James McParland

With Simpkins gone, Orchard moved in for the kill. He rented a room in the private residence of Mrs. W. H. Schenck. His room was a front one that conveniently overlooked the route the victim used in his walks downtown. It proved to be a good observation post.

Jack Simpkins, in the meantime, had gone to Silver City and talked with Bill Easterly, the Cripple Creek derelict who had settled in the camp and was working at the Blaine mine. He told Easterly that Orchard was in Caldwell, and he in turn mentioned the fact to Charles Moyer, who was there on Federation business. The union president sucked hard at the news and wasted little time in getting out of Idaho.

From Silver City Simpkins went to Hailey, another mining community. He intended to stop at Caldwell long enough to give Orchard a letter and some badly needed encouragement. The union official urged the depressed dynamiter not to become discouraged and leave before he had finished Steunenberg. Trying to bolster Orchard's ego and spirits, he mentioned that once the job was finished he would come into a great deal of money, enough to buy a ranch so that he could quit this line of work. The mention of money was the key needed to spur Orchard. In appreciation he handed Simpkins his own train ticket, good from Spokane to Denver. They shook hands and parted.

A few days after Simpkins' departure, Orchard hurried to Boise when he heard that the former governor might be in the city. On November 22, 1905, still using the name Thomas Hogan, he registered at the Idan-ha Hotel and was given room 42. Seeing nothing of his target in the capital city, moods of depression and loneliness overcame him. Needing someone to feed his confidence, he decided to find a partner, someone like Adams or Pettibone. Simpkins had mentioned to Orchard that Easterly was in Silver City. The killer telephoned him and asked if he wanted to take part in the contract. Easterly replied it was impossible for him to leave Silver City. The next day Orchard restated his propositions in a letter, and again Easterly responded that he had a steady job.

Losing his nerve to maneuver alone, Orchard decided on a trip to Salt Lake City. Perhaps he could convince his friend, Charlie Shoddy, to lend him a hand; after all, it was a chance to make money without working for it. At eleven o'clock on November 25, the frustrated killer arrived in Salt Lake and checked into the Cullen Hotel. Upon arriving in Salt Lake City, Orchard had immediately checked at the Siegal Brothers store, the mine owners for whom Shoddy worked. They implied he might be at the mines in Siegal, Nevada; Orchard wrote Shoddy, but was unable to make contact.

Harry Orchard was in Salt Lake three weeks. While he was there, he received a letter from George Pettibone, informing him that John Neville had died suddenly in Goldfield, Nevada.[60] Orchard, seeing an opportunity to use Neville's death to his own advantage, wrote and asked for money, claiming that he had sent Shoddy to Goldfield to murder the saloon keeper. Using basically the same story, he also wrote Simpkins, extending the lie by saying that Shoddy was with him in Salt Lake and that he planned to use him as a partner in the Steunenberg affair. It was a good story, but it failed to secure him the financial support he craved.

Orchard ran into another old acquaintance, Arthur Doolan, who had gone to Colorado with Orchard in 1902. Their final parting had not been a friendly one, so now they were hesitant to shake hands; each waited for the other to speak. Finally, Doolan broke the strained situation, saying: "Heard you got married, Harry?" Orchard grinned, and they walked to the nearest saloon for a drink.

Orchard told Doolan he was selling mining stock for a Denver firm and was expecting company orders that would send him to Spokane on business. He stayed in Salt Lake until December 14, and with the possible exception of a day or two, he spent part of every day with his friend Doolan. Doolan was out of work, so he was available to satisfy the whims of the older man. Orchard was invited to Doolan's house for Thanksgiving dinner, and to show his appreciation, he went to the depot, took out his shotgun and rubber boots from the big trunk, and went out and shot the ducks they had for dinner. Most of their time, though, was spent in the card and billiard room of the Cullen Hotel, where the killer liked to play slough for the drinks.

When Harry left for Caldwell, as in the case of his San Francisco experience, he left incriminating evidence of his profession behind. Several bottles containing sulphuric acid were left on the bureau in his room of the Cullen Hotel. Two times now he had made the mistake of not cleaning up his room properly before he left; the third time might prove fatal. A professional murderer could not afford to be that careless.

Returning to Caldwell, he again registered at the Saratoga Hotel and was given the key to room 9. There was a letter from Pettibone waiting for him. Soon afterwards he received one from Jack Simpkins, enclosing a portion of a typewritten letter from Haywood, which stated that he thought if there were any more remittances for assessment that they had better be sent through him. This is a good example of Haywood's bookkeeping methods regarding illegal expenditures and their justification on the books. In spite of all the talk about his receiving money, the killer still received nothing, and what he had was fast disappearing.

One thing the murderer was successful in accomplishing while he was staying at the Saratoga was cultivating the friendships of certain hotel employees — namely, a young dining-room waitress, Lizzie Voiburg. Orchard generally ate two meals daily, breakfast at noon and dinner at six. Usually, he had a good appetite and enjoyed joking with the waitresses who were serving him, especially Miss Voiburg. He seemed attracted to her and she, to him. Orchard played the role of the perfect gentleman, and a good conversationalist. Before long, they were taking evening strolls when she was off duty.

Several days before the day of the murder, however, he began to change. Becoming sullen and morose, Orchard stopped smiling and appeared nervous at intervals. When Miss Voiburg asked what was troubling him, he replied, "I'm a bad man. Sometime, perhaps, you'll learn how wicked I am, and then you'll loathe and despise me." The killer could have been just trying to impress the woman or merely being dramatic, but later events showed that he would have been better off if he had kept his mouth shut.

The assassin was still frustrated in his attempt to locate the victim, and it was not until Christmas Day that he spotted Steunenberg walking to his brother's house. Getting his pump shotgun, he broke it into two parts, hung it around his neck, and put on his overcoat. He arranged the weapon until it felt comfortable, then walked to a spot close to the victim's house and waited. Becoming chilly, he put the collar of the coat close to his ears for warmth and began moving his feet. Some time later when he heard the former governor approaching, the killer hurried to remove the murder weapon from under his overcoat and assemble it. Hampered by the cumbersom clothing and the cold, he had trouble getting it ready. When the gun was finally loaded and ready to use, the victim had passed and was entering his home.

Swearing, Orchard walked completely around the house, desperate for a close shot. He saw some indistinguishable movements inside, but little else. The assassin waited under a tree in the yard until it was evident that the ex-governor had retired, then gave up the cause for the night. He buried the shotgun shells under a nearby sidewalk, not wanting the evidence in his room if he eventually used the weapon.

That night a masked ball was held at the Saratoga. It was a perfect occasion for the commission of a murder. The former governor could be slain, and in the resulting confusion the killer would not be detected among the other guests. Orchard could then ease upstairs to his room unobserved. It was a plan worthy of consideration; unfortunately, it lacked only one ingredient — a victim. Steunenberg never appeared at the masked ball!

Orchard was unable to learn of the former governor's whereabouts until the next Thursday. Seeing the victim's son, twenty year old Julian, on the street, he boldly stopped him and asked him when his father would be home, explaining he had an appointment with the elder Steunenberg to discuss buying some sheep to feed. He was informed that the boy's father would be in town the coming weekend. On Saturday morning, Julian questioned his father about Orchard, and Steunenberg replied he did not know such a man, and had no such business appointment. He promptly forgot the matter.

On Friday, the determined dynamiter went to Nampa, thinking the victim might be on the train that would come from Boise. Since the train had a stopover in Nampa, he could board it at that time and ride with Steunenberg to its destination at Caldwell. Orchard had placed his bomb in a small satchel, which made it ideal for handling. When the auspicious moment presented itself, the satchel could be placed under Steunenberg's seat without attracting attention. To hell with the other persons on the train. It was unfortunate they had to be at the wrong place at the wrong time.

Orchard described the explosive device that was tailor-made for his victim:

> I had taken the powder out of the wooden box (the original container as fabricated by Orchard and Simpkins) and packed it in a little light sheet iron box with a lock on it and I had a hole cut in the top of this and the little alarm clock and the bottle of acid set in plaster of paris and the giant caps set in the powder underneath this hole in the top of the box, and all I had to do was wind up the alarm and set it.[61]

The destructive device was then fitted into the small inconspicuous grip. When the train arrived in Nampa from Boise, Orchard boarded it and walked through the various cars searching out his victim, without success. Dejectedly, the killer waited morosely until the train pulled into Caldwell. Leaving his place on the train, he headed for the depot and chanced to spot the former chief executive also leaving the train. Failure again! Steunenberg had just bought another day of life, but Orchard vowed it would be his last reprieve.

Time was running out for the man who had by now incurred the wrath of the Federation. This time, Orchard was more than ready to get the Steunenberg killing over, for he remarked: "Well, I got disgusted with myself at not being able to get him, so

I finally decided to use the bomb and get rid of the thing and get away."

On Saturday afternoon, December 30, 1905, Harry Orchard was passing the time away playing cards in the Saratoga saloon. At dusk he stepped out into the hotel lobby to stretch. When he turned, there sat Steunenberg. It was perfect. Taken aback by the suddenness of the situation, the killer walked to the post office, mailed a letter, and weighed the opportunity confronting him. Upon returning, he found Steunenberg still in the hotel lobby. In his pocket was a life insurance policy for $2,000, issued by the New York Life Insurance Company. The ink on it was hardly dry since it had been issued that very day.

Elated with his unexpected opportunity, the assassin hurried to room 19, removed the bomb from the grip, wrapped it in a newspaper as a disguise, and securing it under his right arm, grimly walked downstairs. This was it — the months of waiting would soon be over. Passing through the lobby, he again noticed the victim. Orchard quickly cut through the light snow to the death scene — Steunenberg's house.

Arriving at the familiar west gate leading to the back door of the home, he knelt in the snow and laid his explosive device close to the left gate post. He had discarded the idea of detonating the bomb by an alarm clock, so he took the precaution of fastening fish line, a piece of the same cord used on the Goddard bomb, to the cork of the acid bottle. The free end of the line was now tied around a picket in the gate. When the gate was opened, the pull would eject the cork from the bottle and allow the acid to flow out and detonate the dynamite. Ingeniously, this time the killer had arranged the bomb so that if the gate were not opened wide enough or with enough force to unplug the cork, the victim would trip over the fish line and unwittingly achieve the same effect.

Adding the finishing touches to the booby trap, the killer covered the bomb with the newspaper and then piled snow over it. The death trap was in place, and now Orchard was eager to return to the relative safety of the hotel, where he hoped to provide an alibi for himself. Orchard had been playing the whole bit by ear tonight. Everything had happened so fast he had to rely on instinct now.

Ironically, the assassin met his victim about a block and a half from the death trap. The killer nodded to Steunenberg, a sheer reflex. Then he broke into a run, terrified by the unplanned meeting. There was nobody on the street until he reached a little church were he met two men, one of whom was Julian Steunenberg and the other his uncle. Rattled thoroughly by now, the murderer had intended to cross a bridge at this point, but instead he went straight to the next street and crossed a wooden footbridge that ran along side the Saratoga Hotel. As he was crossing the bridge, the explosion reverberated through the stillness of the evening air. The killer's heart stopped beating for a long second, then it pounded so furiously he was breathless. It was 6:45 p.m. As planned, Steunenberg had opened the gate, stepped inside, and was in the act of closing the gate when the mass of dynamite exploded, catching the former governor on his right side and literally tearing his body apart as it flew ten feet through space.

The forty-four year old victim lay dying on the cold, hard ground, his blood flowing onto the white snow. For the radical members of the Western Federation of Miners, the troubles of the violent Coeur d'Alenes had finally been avenged. The symbol of "armed might" that had ended the ambitions of a young union in the narrow canyons of northern Idaho, was finishing his tenure on earth the hard way.

Reaching the hotel three minutes after the blast report, Orchard hurried directly to the bar; it beckoned as a haven of light. He spoke to the bartender, R.J. Inman, and made sure the man would remember him. The room was empty; the bartender was busily wrapping a package, and he asked the killer to assist him. The slayer appeared steady and composed, hardly like an individual who had just committed murder. After a brief comment on what the mysterious explosion could have been, he had a drink, then went to his room.

At this point Harry Orchard, alias Thomas Hogan, planned to eliminate any evidence that might incriminate him. His first act was to clean off the cayenne pepper he had put on his shoes in case dogs were used to track the killer and then to empty some sulphuric acid in the sink; he put the glass bottle in the right pocket of his coat. A sharp flash and a report much like a pistol

shot occurred. The killer had previously dropped an extra blasting cap in his pocket, and when he slipped the glass vial in, a drop of acid came in contact with the cap, detonating it. The coat was shredded by the explosion, but Orchard remained unharmed.

He was frantic and fully expected to be surrounded by people drawn by the noise, but nobody seemed curious, even though the hotel's dining room was directly underneath room 19. The people below were eating and too engrossed in discussing the Steunenberg bombing. Nonetheless, the accidental firing of the blasting cap completely unnerved the professional killer. "I seemed to lose my reasoning power," he later confessed. For even though he had left bits of plaster of paris, chloride or potash, sugar, bottles, screw-eyes, and tiny pieces of dynamite scattered around the room, the undeniable makings of a bomb, he could not force himself to remain in the confines of the room. He had to get out.

Putting on another coat, he casually walked downstairs and tried to appear natural. Strolling into the dining-room, he was a little pale-faced. When Lizzie Voiburg approached his table, he did not smile at her, and she suspected something was wrong. The killer ate very little, then left the table. Miss Voiburg thought he had just stepped out of the room for a moment and would return, but the murderer had no appetite for any food.

The victim, Frank Steunenberg, fourth governor of the State of Idaho, had ceased living at 7:10 p.m. while his assassin was overcome with a sense of relief and pride in his accomplishment. This was the true highpoint of his professional career.

Within two hours after the news of the Steunenberg assassination reached Boise, a special train was dispatched by the Oregon Short Line to the scene of the murder. Aboard was the present governor of the state, Frank Gooding, the prominent attorney W.E. Borah, former Lt. Governor Joseph H. Hutchinson, a former attorney general of the state, a deputy United States marshal, and several other lawmen and interested individuals. They made the trip in a record time of forty minutes, intent on personally bringing the murderer to justice. There had been nothing like it in the history of the state.

By ten o'clock the Saratoga Hotel was jammed with excited

people. The special train from Boise had arrived, and Caldwell was alive with rumors and activity. The town itself had been sealed tight, surrounded by guards. Roads and other approaches had been closed. Governor Gooding had acted fast to close the avenues of escape.[62]

Amidst the frenzy that accompanied the tragic death of an important person, the man called Tom Hogan sat in the lobby of the hotel and appeared undisturbed by all the talk about murder. Among a group of local persons he recognized a member of the county board of commissioners. He beckoned to the man and asked where he could get a band of wethers cheap. What kind of ridiculous question was that at this particular time? The disgusted man left abruptly to join his friends.

About twelve that night, the "inquisitive" Harry Orchard and the hotel desk clerk strolled down to the murder site. The area was a mess; blood and gore and the wreckage made the slaying believable. The killer noticed that the gate was gone, and a small cone shaped hole eight or ten inches deep in the hard earth gave evidence of the bomb's location. Shreds of the victim's clothing lay scattered over a fifty yard tract. All the windows on the west side of the Steunenberg residence were broken, and glass was shattered in homes as far away as a block. Small pieces of iron, brass, and copper had been blown in every direction.

Immediately after the explosion, the former governor had been taken into his house by neighbors and laid upon a bed. The victim's features were not marred, but his lower body was horribly mangled, and his right arm was nearly torn off. When laid on the bed Steunenberg asked, "Who shot me?" A little later he attempted to sit up, then fell back with the words, "For God's sake, turn me over; I'm dying." The doomed man retained a measure of consciousness and was asked numerous questions, but to no avail. He had suffered extensive hearing damage and could not understand what was being said. Twenty-five minutes after the box of dynamite detonated, he died. The state of Idaho had its first martyr.

Harry Orchard returned to his hotel room and to bed. He slept well — the waiting was over. He slept soundly and awoke with some effort at eleven. Before having his breakfast, he picked up a copy of the Boise newspaper, the *Idaho Daily Statesman*. It carried the date December 31, 1905. The black bold headlines jumped from the page: EX-GOVERNOR STEUNENBERG FALLS VICTIM TO DYNAMITERS! Reading further into the article he received a jolt when he came across the following statement:

> Many people believe it is a result of his firm stand for law and order at the time of the Reign of Terror in the Coeur d'Alenes, in 1899, and the ex-governor has fallen victim to the lawless element that ruled there.

That Steunenberg had been threatened over the years was public knowledge, but most of the threats had been the "crackpot" variety. By January 1, 1906, the reading public was exposed to the hypothesis that an "Inner Circle" of the Western Federation of Miners was responsible for liquidating the former governor. Things were moving fast; the net of the state government was closing in on the killer. Attorney W.E. Borah was getting an early start on Harry Orchard's future. He was with Governor Gooding in Caldwell that very moment and was searching for a murderer.

CHAPTER 21.

I had never before been arrested.
. . . Harry Orchard

Orchard was concerned about the material still in room 19. As he read the paper, he knew he had to get rid of it. In the room after dinner he forced himself to destroy some letters and papers, but for some strange reason he was "afraid to start to clean my room." The killer was aware that a length of fish line had been found and was believed to have been connected with the destructive device. He forced himself to take the remaining fish line off the reel, break it up, and flush it down the toilet, but that was the extent of the clean-up. For some unexplained reason, he just quit. It has been suggested that the mass slayer subconsciously wanted to be caught and punished for his crimes. Maybe he did, but he also reacted like a man unbalanced by a situation over which he was fast losing control.

The killer grew excited, probably the most excited he ever had been in his life. "I had always said that I would not be taken," Orchard said, but he could not bear to touch the weapon that was in his grip. Actually he reacted much in the same manner he had done when Moyer had been arrested — passively.

Out of morbid curiosity, Orchard took one more walk to the victim's home. He went with a local resident, Pat O'Connor, who often played Solo with him at the Saratoga bar. Orchard was calm now, and he even talked of the affair by saying it was a terrible thing and that an American would never do it. It must have been done by a mad Russian or an Italian. O'Connor nodded in agreement and commented that every stranger in Caldwell would soon have to give an account of himself to a Citizens' Committee that was being set up to investigate the crime. The remark began to prey on the killer's mind.

Returning from the walk, Orchard sat in a chair in the lobby and gazed out the window without changing his position. Then he went to the hotel bar and began drinking heavily, something he had not done in Caldwell up to this time. He was plagued by O'Connor's statement about every stranger having to give an account of himself. A man approached Orchard and interrupted his thoughts by pretending to be acquainted with him. The stranger was Sheriff Harvey K. Brown of Baker County, Oregon, who was in Boise on business when word reached that city of the murder. Brown had been part of the entourage that had left Boise Saturday night on the special train to Caldwell.

Sheriff Brown thought he recognized the man, but he was not certain. His real name certainly was not Hogan, that much he knew. Orchard protested that the sheriff was mistaken in his identification, but Brown was adamant on the point. Before leaving, he informed Orchard that he was suspected of having been involved in the assassination. The killer retorted that he would find the local sheriff and straighten the whole thing out, which he proceeded to do.

Orchard located Sheriff Moseley of Ada County and admitted he was a stranger. If the officers wanted to question him or had any suspicions about him, he would submit to an examination at once. So frank and natural was the request that the law officers made but a brief but perfunctory investigation and warned him that he would still have to appear before the Citizens' Committee later in the day.

During the routine investigation of all strangers, the law officers approached Orchard's room to search it. Lizzie Voiburg had also tipped off the officers that a search of the room might be profitable. Upon opening the door, they stared in disbelief, then excitedly proceeded to gather evidence which looked more than circumstantial. Towels were tied around the doorknobs to prevent anyone from peering through the keyhole. Small pieces of plaster of paris and a pinch of explosive powder were found, and a length of fish line was lying near the commode.[63] This unexpected discovery had all the earmarks of being something big!

In the afternoon, the killer appeared before the Citizens' Committee. The informal examination was conducted in the office of Judge Frank J. Smith, and among the committee members was W.E. Borah. The killer seemed to have everything under control, for he responded to the various questions in a frank and natural manner. Orchard admitted being from Denver, explaining that he had a contract with the Mutual Life Insurance Company as their representative, but had not been working with insurance for several months. When questioned about his specific business in Caldwell, he hesitated for the first time, but finally answered that he was in town to buy land, despite the fact that he had repeatedly posed as a sheep buyer.

After the examination was completed, Orchard was interviewed by an enterprising Boise reporter who found the suspect in an amiable mood.

"I can't blame the officers for hauling me up for a quizzing bee. It's probably their duty, but it's rather humiliating. They didn't seem to take stock on my story, probably because I didn't have the goods to prove all the facts, but I guess they'll be satisfied when I show them the evidence. I've got a working contract with the Mutual Life Insurance Company out of the Denver office."

"Are you working for that company here?" pursued the reporter.

"No," Orchard replied, "not exactly." He paused to bite off the end of a fresh cigar and light it. "In fact, I haven't been working for the insurance people for about seven months."

"What has been your business in this part of the country then?" shot the reporter.

"Well, not much of anything. I've been looking for investments in real estate a little." The killer paused, then smiled at the newsman as he contemplated his next statement. "That was a great joke, wasn't it? My being pulled in as a suspected murderer? I can imagine what kind of roasts my friends will hand me when I get back to Denver, if the story ever gets out."

When Orchard had been released after the examination, he was warned not to leave Caldwell. He was tempted to leave that very night, but he knew he was being watched. Even if he succeeded in getting away from town, it was bitter cold outside, and the ground was covered with snow. He very much doubted that he could effect a successful escape. He had to stay and play out his hand. And the stakes were high — his life. The killer sought refuge in the Saratoga bar and in liquor. All at once, Harry Orchard alias Thomas Hogan was not as sure of himself as he pretended to be.

About four o'clock on the afternoon of January 1, 1906, three days after Steunenberg was blown into eternity, Orchard was arrested on the charge of murder. He was taken to the prosecuting attorney's office, searched, and the warrant of arrest was read to him. It was the beginning of the end for Orchard. Taken by surprise, the slayer of eighteen men lamented, "I had never before been arrested."

Julian Steunenberg, the son of the ex-governor, identified the man who called himself Hogan as the man who had asked him questions about the movements of his father. John Rice, a local man, identified the killer as a man seen frequently in the vicinity of the victim's house.

The following day, on Tuesday, the killer was arraigned before Probate Judge M.L. Church of Canyon County. Orchard was well dressed; he wore a neat suit, derby hat, and well-polished shoes. The formal complaint was read to him. His nervousness was evident as he kept his gray eyes on the judge. After the reading, Judge Church stated that as the charge was murder, the prisoner would have to be committed to the charge of the Sheriff, J.C. Nichols. He also informed the defendant that a preliminary hearing would be held as soon as the prisoner had secured counsel and was prepared to go on with the case. "How soon can you get counsel?" he asked.

"If this is published in the papers," replied Orchard, "one lawyer, maybe two, will start at once."

"How long will it take for them to get here?" asked the judge.

"It will take a day and a half from Spokane," answered the prisoner.

He was then informed that he could telegraph for attorneys. At this point, the proceedings were closed, and Orchard was marched back to jail with two officers in front and two officers behind. As the killer scanned the room, his demeanor suggested alarm at his position. He was no longer a casual, disinterested spectator.

Orchard's luggage was discovered in the baggage room of the train depot. The keys had been found in Orchard's shoe. The luggage had never been moved to the hotel although it was less than a block away. When law officers learned that Orchard had visited the baggage room on the day of the murder, their interest was whetted even more. They found a suitcase, a small grip, and a trunk. The suitcase revealed small amounts of dynamite, also some odds and ends of clothing and medicine. The smaller grip contained a Colt automatic pistol and cartridges, a pair of brass knuckles, an electric flashlight, a pair of field glasses, a fishing reel with the line removed, a razor, a small amount of white powder which proved to be an explosive by placing a minute particle of it on iron and striking it with a hammer, and other miscellaneous property.

All in all, it was an important find. Some of the clothing had plainly been used in mining work. A pair of crimpers used to cut fuse and set caps was covered with a whitish material. Part of a kit of burglar tools included an elaborate and extensive instrument used in determining the combination of safes. A shotgun, some candles, and scraps used for climbing trees rounded out the highly suspicious inventory.

In Denver, Moyer released a press dispatch:

> An inquiry into the assassination of former governor Steunenberg of Idaho will be conducted by the Western Federation of Miners. We will do this not only to prepare ourselves against any charges that may be brought against the Federation but also in order to ascertain, if possible, whether or not a member of our organization committed the crime. The affair is to be lamented. No one is more sorry for its perpetration than are the officials of the Federation.[64]

The Inner Circle was obviously becoming nervous. Would Orchard crack? The tension at Federation headquarters was building. Moyer was more affected by the turn of events than was Haywood, although the big man was very much aware of the gravity of the situation. Moyer had spent much time in jail, and the thought of spending any more was horrendous.

Meanwhile, in Butte, Montana, Peter Breen, that former Coeur d'Alene dynamiter of 1892, now a county attorney, let loose his personal feelings on the subject to a reporter. Said Breen:

> What do I know about Steunenberg? You can't make a dead saint out of a live scoundrel, and Frank Steunenberg

> was as loathsome a reptile as ever traveled the earth. No, I don't think the miners killed him, but, for my part, I have no regret at his death and the mystery to me is not that he was killed but that it didn't happen sooner. Why profess to be shocked at it? To kill tyrants is a worthy deed, so why be surprised that this man got his deserts. I was through the labor trouble in Idaho, I have seen his work. I believe Steunenberg was killed by some crank whose weak brain was turned by stories of Steunenberg's outrages.

Breen was not only releasing personal pent-up hate, but also expressing the sentiments of many radical Federation members. These statements attracted attention — no wonder the former dynamiter would become further involved in the case.

A letter from Denver was delivered to Orchard either the third or fourth of January. Sheriff Nichols copied the letter and postmark before giving it to the prisoner. The murderer immediately knew who the writer was, although there was no signature on the letter itself; he recognized Pettibone's handwriting.

> December 30
>
> Your letter received. That was sent to Jack December 21 for you. He should send it so that you ought to have it by this time. Will not write any more this time. Write me as soon as you get to your new field.

The letter referred to money sent to Jack Simpkins who in turn was to get it to Orchard. The strange twist to the whole affair was that Simpkins was using the money mentioned to obtain legal aid for the murderer, about the same time he was reading about it.

The day following Orchard's arraignment for murder, a telegram reached him from the Spokane law firm of Robertson, Miller & Rosenhaupt, the same firm which had represented the Federation interests in the Coeur d'Alenes. It read: "Attorney Fred Miller will start for Caldwell in the morning." The killer gloated; he had been told that if he ran into trouble, he was to keep his mouth shut, and legal assistance would be sent as fast as possible.

Simpkins did as he had promised and arranged for the firm to defend Orchard. Fred Miller started out for Caldwell, but he left the train at Walla Walla. He attributed the reason to illness, but mystery shrouded the affair. Orchard became depressed, and Jack Simpkins sent a cryptic telegram to Federation headquaters which said:

"Cannot get lawyer to defend Hogan. Answer."

Haywood saw a desperate situation developing and realized his number one hatchetman needed the best legal aid. If Orchard thought he had been deserted, he might involve them all. The big man wired R.J. Hanlon, financial secretary of the miners' union at Silver City, Idaho:

> Press dispatches indicate there is another conspiracy to connect the Western Federation of Miners with grave crimes. Several persons in Caldwell have been arrested in pursuance of the conspiracy. The Western Federation of Miners defends no persons guilty of a crime; but in the past it has been found that every one of its members accused of crimes were innocent, and they would have been victims of conspiracy had the organization not aided in defense. Have Mr. Nugent take up the defense of any member of that organization, so that, if innocent, he may be discharged.

John F. Nugent, mentioned in the telegram, was an attorney at Silver City and a personal friend of Haywood during his mining days in that town.[65] Hanlon, though, realized this request placed the local union in a precarious position. His union enjoyed a good labor-management relationship and had not been plagued with serious problems as had other mining districts. Hanlon wanted it to stay that way, so for the time being he refused to consult Nugent. He suspected that hot-headed Federation men in the northern part of the state might be mixed up in the assassination, and if the man were guilty, then the law should be allowed to take its course.

The headquarters of the Western Federation of Miners

became frantic, and with good reason. Sheriff Edward Bell of Teller County, Colorado, had arrived in Caldwell and made positive identification of the prisoner. The man who called himself Thomas Hogan was actually Harry Orchard, wanted in Cripple Creek for blowing up the Independence depot with dynamite and killing thirteen persons.[66] For Harry Orchard the masquerade was over. It was the beginning of the end for the mass murderer. Late in the afternoon after Bell's identification had been made, for the first time Orchard admitted his real name when questioned by Captain Swain in the presence of other law officers:

Q. What is your right name, Thomas Hogan?
A. No sir.

Q. What is it?
A. Harry Orchard.

Satisfied, Swain pushed on with his questioning.

Q. Where were you raised?
A. In New York City, I was born.

CHAPTER 22.

Orchard is not guilty of this crime, and I do not think it will be a very difficult task to prove his innocence.
. . . Fred Miller

This was where he had told his second wife he was born. Swain continued with his interrogation, and the killer admitted he had lived in the Coeur d'Alenes and was acquainted with D.C. Coates, Paulson, and others. He also admitted having been a member of the Western Federation of Miners while he was living in Burke, but added that he was not a member at the present time. Orchard, under fire, also stated that he was acquainted with both Moyer and Haywood, an answer that no doubt raised many eyebrows, including those of Governor Gooding. Swain continued:

Q. Let me ask you this: What explanation have to to make of your conduct and actions in this town? What has been your business?

A. I have not been doing anything; I have been looking for a place. I have been thinking of locating in this part of the county, I have just been loafing here, *I have a private disease*, I have been waiting here to kind of get rid of it. I thought sure of getting a place here; I don't know whether I will or not. I was looking at it . . .

Later that night the prisoner was interrogated again. This time the questioning centered around the incriminating evidence found in his room and trunk:

Q. What did you make in your room?

A. I don't know I made anything except some dice, I made some, or I tried to make some.

Q. What did you make them out of?

A. Plaster of paris.

Q. Were you trying to make loaded dice?
A. Yes, sir.

Q. Where were you the night of the explosion?
A. Between the post-office and the hotel, near the corner of the bank, when the explosion took place.

Q. What did you do then?
A. I went into the saloon, the Saratoga.

Q. What did you do with all this artillery there, those guns?
A. I always carry a six-shooter and a shot-gun.

Q. What about the knuckles?
A. I carry them, I had them in San Francisco when I was there.

Harry Orchard was desperately trying to cover his tracks at this point, but the evidence was fast piling up against him.

On January 6, James J. Sullivan, the Colorado attorney and labor leader, made a surprise visit to Orchard. For the satisfaction of the press, Sullivan claimed he was on his way to Baker City, Oregon, on business and decided "to see if I could identify this man Hogan or Orchard as a man with whom I did business of a civil nature last spring." In the presence of law officers, Sullivan visited the mass slayer:

"Hello, Tom, what are you doing here?" he asked.

"I was arrested for the assassination of Frank Steunenberg. I was going to telegraph you if you hadn't shown up today," Orchard told him.

Sullivan told the killer he had no intention of taking up the case and advised him to employ local attorneys. The conversation could go no deeper because law officers were present.

The Colorado attorney then continued to Baker City, Oregon, where he met a young woman stenographer who was employed in Haywood's Denver office. She accompanied Sullivan as far as Caldwell, then continued east to Denver where she made a full report to Haywood on the matter. Sullivan, in the meantime, stayed in Caldwell long enough to assure the killer he would be afforded legal counsel and to further size up the case. At this time, the attorney probably suggested that it would look much better if Orchard personally contacted the Spokane legal staff and asked for help.

After Sullivan left, Orchard wrote to the Robertson, Miller & Rosenhaupt law firm, asking them to defend him. He received an answer from Fred Miller that he would again start for Caldwell on the first train. The response had an immediate effect on the prisoner. His depression vanished, and on that particular morning his jailer found him singing and even smiling.

Miller arrived as promised. At nine o'clock, he visited the jail and asked that the prisoner be advised of his presence, but that he was unable to interview him until after lunch. When the press caught up with the lawyer, he announced that he was going to defend Orchard and explained that the first time he had started out for Caldwell he had to leave the train at Walla Walla because of recurring attacks of neuralgia.

When Miller saw Orchard, he confirmed that Jack Simpkins had sent him and that he had traveled as far as Walla Walla when he received word that he was to proceed no further as the newspapers were already charging the Western Federation of Miners with the former governor's death. He stressed that Federation leadership was most anxious that it not appear as if it were directly involved in providing legal assistance to the accused assassin. They wanted it to seem that Orchard had made the first overtures and was supplying his own defense.

Miller also said that before he left Spokane, his partner Robertson had suggested to make it seem they were engaged by Harry Orchard to sue Don Cordonia to recover the interest he had sold him in the Hercules mine. It was important to make it appear that they were his regular attorneys. He mentioned that Simpkins had given him one hundred dollars as a retainer to handle the case.[67] Orchard told Miller to get in touch with James Sullivan and have him forward the defense money that he expected from Denver. He also told Miller to contact his old friend, Lewis Cutler, of Salt Lake City. The man would turn over a sixth interest in the mining claims he had in Goldfield. Orchard had loaned Cutler a little money at different times, and to repay the debt, he had given the killer his interest in mining claims.

The preliminary examination of Harry Orchard was held in Probate Judge M.L. Church's office at ten twenty-five o'clock, on January 13th. The prisoner appeared calm throughout the proceedings, and occasionally a smile of amusement or sarcasm crossed his face. He sat close to attorney Miller, and the two had frequent, whispered conversations.

Dr. J.W. Glue testified that the victim died of concussion and shock. Joseph Hutchinson reported finding the fish line at the murder scene, and the prosecution declared this essential evidence. "I was also there at the time a small piece of fish line was picked up. The fish line was of waxed silk about five feet long."[68]

After the noon recess, Attorney James H. Hawley, who had been retained as chief prosecutor for the State of Idaho, raised an objection to the presence of reporters. Perhaps he recognized even at this early date the significance of the case and the national prominence it would achieve. He stated firmly:

> It is not to the interest of the prosecution to have any stories written about what transpires here. We have a hard case before us. Detailed reports of these proceedings or summaries of the evidence taken here cannot help us in prosecuting this man. The very men who will be apt to read these reports are those who may want to compose a jury of intelligent men. They cannot help but form opinions in reading about the evidence submitted that will disqualify them as jurymen.

Hawley was sustained in his objection, and the grumbling reporters were sent out into the cold hall as the hearing continued. Sheriff Bell of Cripple Creek identified the prisoner as Harry Orchard and stated that he had a warrant for his arrest and that extradition papers had been issued for his return to Colorado in the event he was not convicted in the Idaho courts.

Joseph Hutchinson was recalled as a witness since he had also been the first man to gain admittance to the prisoner's room in the Saratoga Hotel. He testified to his findings there. As he spoke, Orchard had assumed an air of disinterest until

Hutchinson described the things which had been found in room 19 and remarked about the towels tied around the doorknobs to cover the keyhole. This testimony noticeably disturbed the defendant, and he whispered excitedly to his lawyer.

Twenty witnesses subsequently were presented by the prosecution. When the defense offered no evidence, Miller explained his reasons for keeping silent at this time: "Orchard is not guilty of this crime, and I do not think it will be a very difficult task to prove his innocence." As proof of his statement, Miller claimed that the instrument found in the prisoner's luggage and said to be a contrivance to determine the combination of safes was nothing of the sort. It was used in making loaded dice. Furthermore, the plaster of paris found in room 19 and in the defendant's trunk was also used to construct the illegal dice, Miller contended.

On January 15, Harry Orchard was bound over to the District Court of Canyon County, without bond, to be tried for the murder of Frank Steunenberg.[69] When the verdict of the court was given, the accused still seemed to be unconcerned by this turn of events. He had lost no weight and had demonstrated repeatedly that his appetite was not hampered by his confinement.

Desiring the very best investigating services available, and at the urging of Chief Justice E.C. Stockslager of the Idaho Supreme Court, Governor Gooding hired the most famous detective agency in America, the Pinkertons. Taking personal charge of the case was a mild looking, nattily-dressed gentleman with an immense walrus moustache and gold rimmed glasses that covered penetrating eyes. With thinning gray hair touched with red, the man looked like someone's grandfather. But he was a man who had become a legend — James McParland. McParland was manager of the western division of the Pinkerton agency, and his office was in Denver.

As a young man in the 1870s, McParland, as an undercover Pinkerton agent, had successfully infiltrated the infamous Molly Maguires, a secret terrorist society in the anthracite coal fields of Pennsylvania. Over a period of years the detective had witnessed some of the most hair-raising feats imaginable and collected enough evidence to break up the murderous group and bring many of its worst members to trial. The accomplishment so impressed A. Conan Doyle that he wrote a novel based on McParland's adventures entitled *Valley of Fear*.

On January 10, McParland arrived in Boise to take over the murder case. During a meeting with the governor and the chief justice, the evidence in the case was covered in detail. After the high powered conference, the detective concluded, "I am satisfied that there were other people in this plot besides Orchard and feel sure that Orchard was the tool of others."

One of McParland's first acts was to get the killer removed from the Caldwell jail to the state penitentiary, on the outskirts of Boise. There were several reasons for such a move. The state institution would be more secure and would help to deter a possible escape try or an assassination attempt on the prisoner himself. Probably the main reason was that the detective wanted to set up a psychological situation that would help him to obtain the needed confession from Orchard. A shrewd judge of men and their behavior, McParland wanted the killer placed in a solitary cell on death row. "Place guards who will watch him day and night, who will never take their eyes off him, either day or night. At the same time never speaking to him or allow him any visitors except his lawyers . . ." McParland instructed the authorities. The biggest drawback to the proposal was district court judge, Frank Smith, who had certain reservations. He feared he might jeopardize the state's case by making an illegal transfer that could result in Orchard's being freed on a writ of habeas corpus. McParland solved the dilemma by persuading Governor Gooding to meet with Justice Stockslager and Sullivan to receive their opinions on the legality of such a move. The meeting took place, and the justice rendered an informal decision that such a move was perfectly legal and justified.

According to McParland's request, Orchard was placed in a cell on death row, where guards checked him all day and night without speaking to him. The prisoner was given no reading material, and at one time Orchard wrote:

> I think I had been at the penitentiary about ten days or two weeks, and the warden took me out into the secretary's office and introduced me to an old man . . .

That old man was McParland, who observed: "Orchard was about as determined a countenance as I have ever seen on a human being, with the most cold cruel eyes I remember having seen; his cheek bones are very high, and he has an extremely large and well-formed forehead . . ."

The two were together on Monday, January 22, 1906, from two in the afternoon until five-thirty. McParland's first overture was at first friendly. Then he tried the scare approach by skillfully playing on his feelings. At the beginning of the inquiry, McParland mentioned that he was a detective, but avoided revealing his true identity. Orchard showed he was willing to talk — he was probably relieved after his silent confinement period — but he made it clear that he was innocent, and that McParland was simply wasting his time with him.

The detective skillfully veered the subject to the involvement of state witnesses and hit hard at cases in which the state witnesses had been set free. McParland told the prisoner he believed him to be just a tool used by the interests of the Inner Circle of the Western Federation of Miners, and stressed the importance of giving state evidence. As an example, the detective told Orchard the story of Kelly the Bum, who, along with several other men, had taken Alexander Ray's watch and all his money. Ray was down on his knees praying for his life while the other men were deliberating on what to do with him since he knew them all. Kelly the Bum settled the discussion by shooting the man in the head, exclaiming that "Dead dogs tell no tales." When the men were brought to trial, Kelly turned state's evidence by admitting that he was the one who had fired the fatal shot. The others were hanged for the crime, while the indictment against Kelly was dismissed.

McParland asked Orchard if he had ever heard of the Molly Maguires. As Orchard later related, "He started to tell me about them, and it struck me right away that it was McParland, as Haywood had given me a description of him some time before." What Haywood said of McParland was that "the old rascal ought to be killed," but he also mentioned that he had never bothered them. "I asked him if his name was not McParland, and he said it was." The killer was duly impressed, but he still asked to see the detective's membership card in the Elks lodge for identification purposes.

One important fact that McParland managed to have Orchard admit was that Jack Simpkins had been with him in Caldwell and Nampa. As the meeting ended, the killer was interested enough to ask McParland to be sure to call on him again. That evening, the exuberant detective met with Governor Gooding, and the two men stayed up until twelve-thirty to discuss the day's activities.

The second meeting of the detective and the murderer was on Wednesday, the 25th. After a brief talk with Warden Whitney, the prisoner was ushered into the warden's private office, and they were left alone. McParland opened the conversation by asking Orchard if he had thoroughly considered the conversation they had on Monday and more especially the suggestion about state's witnesses.

Harry Orchard told the detective that he had thought of little else, and McParland was satisfied that the killer was prepared to plunge into the set trap. Orchard was anxious to know what he could get out of turning state's evidence. McParland made it clear that if Orchard did so, there was no guarantee that immunity from prosecution would be granted. On the other hand, Orchard should not jump to the mistaken conclusion that all state witnesses suffered the supreme penalty. Every objection Orchard suggested to one's offering evidence, the detective countered with a logical and convincing explanation.

In his clever interrogation, McParland left no areas uncovered. His next angle dealt with religious and moral aspects. He asked the prisoner if he believed in an all-seeing and divine Providence. The surprised Orchard replied that he did. Did he believe in the hereafter? McParland pursued. "Ye-es," answered the mass slayer. McParland, the veteran actor, hid his delight at having discovered a new realm to explore, one that would eventually reap dividends. The poker-faced killer, however, was alert to the new implication.

Shifting the line of thought again, McParland now stated that he had not come to discover whether or not the prisoner was guilty of the Steunenberg murder. Positive proof was at hand. What he wanted to know was who hired Orchard? The questions were coming too fast. Orchard, silent, had qualms about exposing the Inner Circle. McParland saw his mistake and so returned to the religious theme. It was an awful thing to die before repenting of one's sins. Did Orchard realize that there was no sin that God would not forgive? The detective then told him the story of King David and the Apostle Paul, murderers who had repented and turned to religion. Every man should repent of his sins! he insisted.

Orchard was weakening. McParland had touched a nerve, and he realized the breaking point was at hand. Once more the detective reiterated that he thought Orchard could escape the hangman. Frustrated, and suddenly overwhelmed, the killer jumped out of his chair and went to McParland, "My God, if I could only place confidence in you! I want to talk . . ." he cried out.

McParland had won. Orchard was ready to confess.

Later, McParland summarized his success by saying, "I found that he prides himself on being very intellectual, and I catered to his vanity in that respect all through the conversation." He had sold himself to the killer. From that day on, he would be father-confessor, counselor, friend, and his only contact with the outside world. In fact, when the prisoner was being returned to his cell, he turned to McParland and said, "Now don't forget to come here tomorrow." As if the detective could be kept away! That night Orchard tried to pray. He could not remember the last time he had done that.

The following day, on Friday, January 27, Harry Orchard began his confession. W.B. Hopkins, private secretary to McParland, recorded as the killer narrated. The dictation continued until the evening of the 31st. Once the flow of words started, an unemotional, calloused confession emerged. The testimony was, nevertheless, an outstanding and revealing document. As Governor Frank Gooding said, it was "so full of horror as to be unbelievable." This story, a story of a man's life in crime, staggered the imagination, but every word of it was true.

Meanwhile, Fred Miller, the attorney hired by the Federation to defend Orchard, was on his way to the state penitentiary with a substantial defense fund at his disposal.

Fig. 16. Governor Frank Steunenberg.

Fig. 17. Site of the fatal bomb blast that assassinated former governor Steunenberg beside his home in Caldwell, Idaho. Note the crater in the ground and the shattered gate and fence.

Fig. 18. Mug shots of Harry Orchard at the time of his arrest in Caldwell, Idaho.

Fig. 19. Steve Adams at the time of the trial.

PINKERTON'S NATIONAL DETECTIVE AGENCY.

FOUNDED BY ALLAN PINKERTON 1850.

ROB'T A. PINKERTON, NEW YORK.
WM. A. PINKERTON, CHICAGO.
PRINCIPALS.

GEO. D BANGS,
GENERAL MANAGER
New York.

ALLAN PINKERTON,
ASS'T GENERAL MANAGER,
New York

—OFFICES.—

EASTERN DIVISION.
JOHN CORNISH, MANAGER,
New York.

NEW YORK	57 Broadway.
BOSTON	30 Court Street.
MONTREAL	Merchants Bank Building.
BUFFALO	Fidelity Building.
PHILADELPHIA	441 Chestnut Street.
PITTSBURGH	Second National Bank Building.
CLEVELAND	Garfield Building.

POLICE

MIDDLE DIVISION.
E. S. GAYLOR, MANAGER,
Chicago.

CHICAGO	201 Fifth Avenue.
ST. PAUL	Manhattan Building.
MINNEAPOLIS	Metropolitan Life Ins. Co. Bldg
KANSAS CITY	622 Main Street.
ST. LOUIS	Wainwright Building.
CINCINNATI	Mercantile Library Building.

MEMBER NATIONAL BUREAU OF IDENTIFICATION WASHINGTON, D.C.

WESTERN DIVISION.
JAS. McPARLAND, MANAGER,
Denver.

DENVER	Opera House Bloc
OMAHA	New York Life Buildin
PORTLAND, ORE.	Marquam Bloc
SEATTLE	Alaska Buildi
SPOKANE	Rockery Buildi
LOS ANGELES	Wilcox Buildi
SAN FRANCISCO	James Flood B

SPOKANE,
ROOKERY BUILDING.
G. J. HASSON, S

ATTORNEYS.
GUTHRIE, CRAVATH & HENDERSON,
New York.

REPRESENTING
THE AMERICAN BANKERS' ASSOCIATION.

$2,000.00 REWARD

Governor Frank R. Gooding, of the State of Idaho, will pay a reward of Two Thousand Dollars for the arrest, or information leading to the arrest of

L. J. SIMPKINS, alias J. SIMMONS

Simpkins, alias Simmons, is charged with the murder of Ex-Governor Frank J Steunenberg, of Idaho, at Caldwell, Idaho, on the evening of December 30th, 1905, and a warrant has been issued for his arrest.

L. J. SIMPKINS

DESCRIPTION:

Name	L. J. Simpkins.
Alias	J. Simmons.
Age	40 years.
Height	5 ft. 8 1-2 or 9 in.
Weight	180 pounds.
Build	Heavy, thick chested, slightly stoop shouldered.
Eyes	Blue, with decided cast, shifty glance, very peculiar look.
Nose	Large, slightly crooked.
Teeth	Large, prominent upper front teeth.
Hair	Dark.
Mustache	Heavy, dark and tawny.
Beard	Dark and heavy; may be smooth faced.
Complexion	Medium.

L. J. SIMPKINS

When last seen wore dark gray coat; gray trousers of cheap appearance; black fedora hat with high crown and large brim; long black overcoat; white lay down collar; medium colored four-in-hand tie.

Is member of the Executive Committee of the Western Federation of Miners.

If located, arrest and notify any of the above listed offices by telegraph or telephone.

Pinkerton's National Detective Agency,
306 Rookery Building, Spokane, Washington.
Day Phone Main 234. Night Phone Main 6647.

Or

G. J. HASSON, Superintendent.

Spokane, Wash., April 5, 1906.

Fig. 20. Award poster for the capture of L. J. Simpkins.

Fig. 21. Harry Orchard being transferred from the prison to the courthouse. (1) Warden E. L. Whitney, (2) Deputy Bartell, witness for the state from Telluride, (3) Harry Orchard, (4) James McParland, (5) Edgar Hawley, and (6) Detective Charles A. Siringo of the Pinkerton Agency.

Fig. 22. Left to right: Mr. and Mrs. George Pettibone, William D. Haywood, and Mrs. and Mr. Charles H. Moyer, at the time of the trial.

Fig. 23. The prosecution at the Haywood trial. Left to right: Senator-elect William E. Borah, W. A. Stone, and James H. Hawley. Standing: Owen M. Van Duyn, district attorney of Canyon County.

Fig. 24. Courtroom scene with Harry Orchard on the witness stand (left arrow) being cross-examined by Edmund F. Richardson of the defense behind the water pitcher on the defense table. Clarence Darrow (right arrow) listens. Relatives, court officials, and gunmen for the two sides fill the rest of the room. The prosecution staff does not appear in this photograph.

Witnesses for the Defense in the Haywood Trial

1. F. C. Clifford, Denver insurance agent. 2. J. Wolff, manager of Pettibone's store. 3. Paul Corcoran, two years in penitentiary for alleged murder in Coeur d'Alenes. Sent home without testifying—a fine fellow.) 4. F. R. Red, Cripple Creek, deported. 5. Henry Macki, tied to telephone pole in Telluride by militia, hands frozen. 6. C. W. Aller, telegraph operator, Cripple Creek. 7. Chas. A. Sullivan, Cripple Creek miner. (Impeaching witness Orchard.) 8. Chas. A. Collins, miner, Independence. 9. George Breon, railroad conductor on train from Victor to Dunville, military kidnapping expedition. 10. Pat Moran, saloon keeper, Cheyenne, Wyo. (Implicated by Orchard.) 11. Will B. Easterly, Cripple Creek, member of W. F. M. (Implicated by Orchard.) 12. John M. O'Neill, editor of "Miners' Magazine." 13. M. E. White, organizer for the W. F. M., Leadville, Colo. 14. George Hooten, union store keeper, Anaconda. (Was in bull-pen.) 15. James A. Baker, ex-member executive board W. F. M. for British Columbia. 16. L. F. Taylor, miner, Independence. 17. John Dennis, Cripple Creek miner. (Escaped deportation.) 18. Tom Foster, Cripple Creek miner. (Deported. Implicated by Orchard.) 19. A. H. Floaten, Telluride business man. (Deported.) 20. Tom Woods, Cripple Creek miner. (Beaten by militia.) 21. Morris Friedman, author of "The Pinkerton Spy." 22. Max Malich, Globeville politician. (Implicated by Orchard.) 23. Mrs. Sadie Cox, Goldfield, Colo., member of Woman's Auxiliary. (Militia victim.) 24. Mrs. Chas. H. Moyer. 25. Mrs. Moyer's sister, Mrs. T. Noggle. 26. Marion Moor, member of executive board of W. F. M. 27. V. C. Barnes, Telluride. (Deported.) 28. Mrs. George Hooton, of the Woman's Auxiliary of Cripple Creek. (Militia Victim.) 29. Mrs. Mark H. Morrison, Cripple Creek boarding house keeper. (Militia victim.) 30. Rev. T. S. Leland, pastor of M. E. Church, Victor. (Was in bull-pen.) 311. W. F. Davis, strike committee, Cripple Creek. (Implicated by Orchard.) 32. Mrs. Steve Adams. 33. Mrs. George Pettibone. 34. Alex. Paul, secretary of Cripple Creek Union No. 40. (Deported.) 35. Mrs. Nellie M. Joyce, of the Victor Woman's Auxiliary. (Militia victim.) 36. Miss Anna M. Saunders, Cripple Creek Woman's Auxiliary. (Militia victim.) 37. Frank Aikens, Cripple Creek engineer. (Was in bull-pen.) 38. D. C. Copley, ex-member of the executive board of W. F. M., Cripple Creek 39. William Amole, Victor, watchman in Portland mine. (Beaten and robbed by mob.) 40. "Owney" Barnes, Cripple Creek district miner. (Implicated by Orchard.) 41. Chas. A. McKinnon, W. D. Haywood's brother-in-law. 42. Joseph Mahalich, Globeville. (Implicated by Orchard.)

Fig. 25. Witnesses for the defense in the Haywood trial.

Fig. 26. Harry Orchard in the 1930s.

Fig. 27. Statue of Governor Steunenberg in Boise. The epitaph reads:

FRANK STEUNENBERG
GOVERNOR OF IDAHO
1897-1900

WHEN IN 1899 ORGANIZED LAWLESSNESS CHALLENGED THE POWER OF IDAHO, HE UPHELD THE DIGNITY OF THE STATE, ENFORCED ITS AUTHORITY AND RESTORED LAW AND ORDER WITHIN ITS BOUNDARIES FOR WHICH HE WAS ASSASSINATED IN 1905. "RUGGED IN BODY, RESOLUTE IN MIND, MASSIVE IN THE STRENGTH OF HIS CONVICTIONS. HE WAS THE GRANITE HEWN." IN GRATEFUL MEMORY OF HIS COURAGEOUS DEVOTION TO PUBLIC DUTY, THE PEOPLE OF IDAHO HAVE ERECTED THIS MONUMENT.

CHAPTER 23.

Here we were in murderer's row, in the penitentiary, arrested without warrant, extradited without warrant, and under death watch!
. . . William Haywood

On Saturday, February 3, Fred Miller arrived to defend Harry Orchard — totally unaware that the man had already confessed. McParland had done everything he could to protect the secrecy of the confession until the members of the Inner Circle could be arrested and were in protective custody. It was, therefore, essential that this interview appear normal. The detective was also interested in what the union officials were doing and what their plans were. This was a rare opportunity indeed, and McParland was scheming to make the most of it.

Miller told his murderous client that he had gone to Denver and had received fifteen hundred dollars for his defense fund.[70] By adding the hundred dollars from Jack Simpkins, it now totaled sixteen hundred dollars. The wily lawyer had also stopped over in Salt Lake City, and Cutler had promised to sign the Nevada mining claims over to his friend, Orchard. All in all it had resulted in a profitable trip, and Miller was elated. The attorney also said he had been authorized to employ a local lawyer to assist with the case. Moyer had also instructed him to buy the prisoner a new suit of clothes and also leave one hundred dollars with the warden, for Orchard's personal use.

Miller had also seen the killer's wife in Cripple Creek and had asked if she had any knowledge that her husband had ever been kicked in the head by a horse? This implied that he might not be in possession of his full mental facilities. The woman replied that Harry certainly did have a scar about an inch behind the hair line of the forehead. Yes, Harry had told her that at one time he had fallen off a horse. Some grey hair now grew over the scar. Miller realized the story might be useful in the courtroom.

The lawyer also pulled another surprise on his client when he stated that he knew why Harry had been in Caldwell and in Nampa. A party had informed the dynamiter that he knew where he could make a little money — more specifically, that some sporting-gentlemen from Nampa and Caldwell were very fond of gambling, especially dice, and that this party had shown Orchard how to make loaded dice out of plaster-of-paris so that he could skin these people. Orchard realized that Miller was searching for a way to explain the presence of plaster-of-paris found in his room at the Saratoga. The attorney went on to mention that he had hired a private detective in Denver who was making an investigation to show that Orchard had a first class reputation.

Miller went on to say that he did not know where Jack Simpkins was, except he was in hiding.[71] Simpkins had told him that Steve Adams was living somewhere in Oregon and could be located if necessary. Miller added that Max Malich would testify at the trial that he had requested Orchard to buy some sheep for his ranch near Montrose, Colorado, and thus provide an alibi for the murderer's assertion that he had been in Caldwell and Boise on business.

To further strengthen the case, Pettibone had promised a deposition stating that he had employed Orchard to construct loaded dice and phony chips, which were made out of plaster-of-paris. Orchard told Miller he wanted Pettibone to appear at the trial in person since he thought the merchant would make a good, impressive witness, and Miller agreed that he would. In actuality Pettibone had no intention of ever appearing at Orchard's trial and had informed Miller of this, although he had no objection to making a legal deposition if it would strengthen the case.

To relieve the mounting suspicions of Haywood, Moyer, and Pettibone and to give them the impression that the state was still groping for clues in the murder investigation, McParland used a ruse. The chief prosecutor, Hawley, leaked an article to the Associated Press, stating that investigative services had discovered that there were two men in Caldwell the night of the murder. One of them was named Simpkins, and detectives were presently on the trail of the two men. The prosecution wished to convey a false feeling of security since the union leaders were well aware that Jack Simpkins was not in Caldwell and that no other person had assisted Orchard in the commission of the crime. The state realized that Miller would immediately telegraph this news to the Denver headquarters.

Later, Harry Orchard read over the original copies of his confession, about 135 pages, and inserted minor corrections as he went. McParland knew only too well that the confession he had obtained was merely the start of a valid case against the conspirators of the Inner Circle. The detective's work narrowed to one principal objective — to corroborate Orchard's testimony. He knew only too well the problems involved in legally proving that a conspiracy actually had existed. Telegrams were sent to Pinkerton's offices throughout the West with specific assignments that would try to corroborate the statements in Orchard's confession. At the same time, the detective agency was checking on information in the case. These reports were extremely important since they helped to dispel the stories that had existed over the years that the Pinkerton Agency was manufacturing evidence which it could then give to Orchard to memorize. The mass slayer was clearly calling the shots.

To safeguard the investigation, McParland drew up a cipher-code of names that was to be used in connection with the case. Steve Adams was referred to as fox, Orchard as possum, McParland as owl, Governor Gooding as lion, Haywood as viper, Moyer as copperhead, Pettibone as rattler; Simpkins was named scorpion, Vincent St. John, coyote, Hawley, tiger; and the Pinkerton National Detective Agency was referred to as justice. Locations also were tagged with a cipher name. Cripple Creek was called Pekin; Burke, Idaho, was known as Frankfort; Denver was Canton; Caldwell was known as Houston, and Boise was referred to as Wheeling.

Miller continued to see his client and reassured him that everything was progressing well, still unaware that Orchard had already turned state's witness. He told the prisoner about interviewing a man registered at the Saratoga Hotel who definitely saw Orchard enter the dining room a few minutes after the explosion. The man believed him to be innocent of the murder and would testify in his behalf. Miller also mentioned that he wanted a continuance in the case until a later date, because he knew that the prosecution was trying to involve the Western Federation of Miners. He wanted to have the case postponed until the June term of the court, which was exactly what the prosecution wanted. It, too, needed more time to build its case.

Orchard had also asked Miller if Edmund F. Richardson, the Denver attorney for the Federation, could represent him. The lawyer told him this had been discussed with the union leaders, but the idea had been vetoed because of the work Richardson had done with the Federation, not to mention that it might be interpreted as a sign that there was a connection between Orchard and the union. Miller was somewhat in the same category, but he had informed the press that he had quit his law firm and had accepted the job as a free agent.

In the afternoon of February 9th, McParland visited the penitentiary and had a long talk with his prize prisoner. Although Orchard had checked the confession for errors and corrections, he had not yet signed his name to it. McParland was becoming nervous. The whole case might rest on the legality of that document, and he wanted all the loose ends tied. When he approached the prisoner about it, Orchard replied, "You will remember when I started in to make this statement, you said it would not be necessary to make an affidavit of it." McParland replied this had been true at that time, but there might be some complications in the matter of extraditing the Federation leaders from Colorado. In all probability it would be necessary to show the governor of that state the actual confession, and it would not be legally recognized if the document were not signed and sworn to, McParland stressed. Orchard pondered the request, recognized its importance, and finally agreed. But before he signed, the prisoner dramatically remarked, "Now, gentlemen, I know there is no trick to this. I look upon Mr. McParland as a father, and I know that he will not forsake me; these statements are true, and I am willing to confirm them." With that, the infamous confession was signed. Despite the promising efforts of his lawyer which might have won him an acquittal, the killer had cast his fate with the State of Idaho.

On Tuesday morning, February 13, McParland met with Goddard and confidentially informed him of Orchard's confession and also of the threats made against the judge's own life. Immediately after the conference, Goddard returned home and examined the gate-frame in his yard. There he found the screw-eye in the wood just as McParland had described it. He examined the ground, and he finally understood why no grass would grow over a certain bare spot. He knew without a doubt where the explosive device was located.

The next day the bomb was removed by the adjutant general of Colorado himself. Bulkeley Wells, like most mining men, was something of an expert with explosives. There was the wooden box, ten pounds of dynamite, and one hundred blasting caps — an unexploded bundle of death. Gingerly, Wells removed the bomb and took it to the laboratory of Von Schultz and Low for examination. It was found that the bomb had been buried for several days during which time the gate had not been opened. Meanwhile, the bottled sulphuric acid had partially disintegrated the head of the pin in the cork, so that the cord connected to the gate was unable to uncork the bottle so that the acid would drip into the box or onto the cotton saturated with potassium chlorate and sugar. The remainder of the pin was secure, sealing the bottle to prevent leakage. Had the disintegration not occurred, the bomb would have gone off.

A demonstration of the deadly bomb was conducted at the laboratory. A small quantity of sugar and potassium chlorate were reduced to a fine powder in a mortar, and then placed on an earthen plate. One drop of sulphuric acid was dropped at the side of the mixture, and in a hundredth of a second the three elements combined, and a violent, colored-flame flared up, fully three feet about the miniature furnace, giving off most intense heat. Had the acid eaten its way through the cork and touched the cotton, nothing on earth could have prevented a terrific explosion.

McParland probably picked this particular way of introducing Orchard's confession to the judge to impress him with the magnitude of his case. The fact was the detective had already planned how he was going to get the Inner Circle criminals out of Colorado. Cooperation from government officials was imperative. The detective had convinced Goddard of the validity of the confession; he was now completely in favor of arresting the Federation leaders and bringing them to Idaho for trial. Goddard also assured McParland that he would inform

Colorado Governor Jesse F. McDonald of the details of the case.

On the sixteenth, there was a high-powered meeting between Goddard, Governor McDonald, Chief Justice Gabbert, James Williams, who represented the interests of several large mine owners, prosecutor James Hawley, and McParland. The conference room was guarded by Adjutant General Wells. The first order of business was a review of Orchard's confession. After hearing it, all men present were in unanimous agreement that the top union leaders — Haywood, Moyer, and Pettibone — should be arrested at once and transported to Idaho to stand trial.

McParland then brought up the customary referral of the extradition papers to the office of the attorney general. Tradition called for the governor to refer the official applications to that state official. McParland was adamant that the attorney general be passed over in the extradition process, for he felt that the man was too heavy a drinker to be dependable. Not only that, but his deputy was a friend of both Moyer and Haywood, and he had been at one time a member of the Western Federation of Miners. If the unorthodox suggestion of the detective were carried out, it would place the governor in a position where he could be crucified by an unfriendly press that could possibly spell the end of his political career. It was an important decision.

Governor McDonald picked up the papers from Hawley, examined them, and announced, "I will sign them, and the record will not go into the secretary of state's office until sometime next week; I hope that the prisoners will then be safely in Idaho."

After this, McParland pointed out that if the arrest of the Federation officials were known, friends of the prisoners might attempt to rescue them, legally or illegally. One legal way open would be to have a writ of habeas corpus served on the officials in charge of the prisoners while they were enroute through Colorado and Wyoming. The illegal method of rescue could be an attempt to forcibly seize the prisoners from their guards. For this reason, it was crucial that a most unusually capable man be in charge of the train carrying the prisoners. McParland recommended that General Wells be detailed to take charge of the train. Wells not only had nerves of steel, but, in case of an emergency, he could operate an engine as well as any locomotive engineer in the state. Governor McDonald agreed and officially authorized him to take charge.[72]

McParland next met with officials of the Union Pacific Railroad to make final arrangements for the special train that would "haul" the prisoners to Idaho. It was to consist of one baggage car, one Pullman car, and a car for the superintendent of the division over which the train would pass. The train would be subject to the orders of General Wells, and the car carrying the prisoners and deputies would be locked. As an added precaution, the special train would not change crews at any division point but would make the change at a telegraph station.

Meanwhile, McParland had kept a constant shadow on the Federation leaders. He planned to have the local sheriff procure three carriages and three dependable drivers. Four deputies, including the driver, were to be placed in each carriage. One Pinkerton agent, who knew the man they wanted, would accompany that particular carriage to identify him. The sheriff's men and the Pinkerton agents were to meet at the courthouse at 6:30 a.m. on Sunday morning, February 18, and from there a carriage would proceed to the residence of each of the three union leaders. From there, the prisoners were to be taken directly to the waiting train and turned over to the custody of General Wells.

Moyer disrupted the carefully made timetable when it was reported to McParland that he was going to leave Denver on Saturday evening on a train bound for the Black Hills of South Dakota. Moyer was quietly arrested aboard the Deadwood sleeper. In his grip was a .44 automatic Colt and one hundred rounds of ammunition. McParland was confident that Moyer's move was an escape attempt.

Haywood was arrested in a rooming house on Fifteenth Street, later to be described as a "house of assignation." Embarrassed, Big Bill was caught unceremoniously in bed with a woman definitely not his wife. The notorious Federation man had been literally caught with his pants down! Lying on the dressing table in the room was a .44 Colt revolver.

George Pettibone was arrested at his home; the three men were taken to the courthouse and assigned separate cells.

Despite the strict secrecy of the arrests, the news still leaked out. Mr. Hawkins of the law firm of Richardson and Hawkins telephoned the county jail to see if the three men were being held there, but was given the "runaround" by the jailer.

About 2 a.m., the three prisoners were removed to the Oxford Hotel, a block from Union Station, and at 5:30 a.m. on Sunday morning, February 18, 1906, they were legally turned over to Deputy Warden J.C. Mills of the Idaho State Penitentiary. At six o'clock, the race for the Idaho border began. Aboard the train, Haywood and Pettibone played cards and talked while Moyer was sullen and silent. He had been through this before.

One place in particular along the route that might have presented a problem was Cheyenne, Wyoming, where a United States marshal was in residence. He had the power to serve federal writs. However, trouble was avoided by not stopping the train at the city and by traveling straight through at thirty miles per hour — too fast for a man to board. In the meantime, McParland telegraphed the Idaho governor the news in cipher:

> Tell Lion (Governor Gooding) excursion train left six a.m. for Wheeling (Boise) with full crew. Morning Republican (Denver Newspaper) claims possum (Orchard) confessed. This comes from state house. Think fox (Steve Adams) talked with coyote (Vincent St. John) on assassination. If so, we can try coyote (St. John) at Caldwell, if fox (Adams) is taken. If possum (Orchard) is accused of confessing, have him stand pat; claim that there is a spy on executive board.

At 9:15 a.m. on Monday morning, February 19th, the special train reached Boise, without incident, and the prisoners were escorted to the security of the Idaho State Penitentiary. The mission had been completed. The prisoners were searched and assigned individual cells. The full impact of what had happened finally caught up with the big man. Haywood wrote:

> Here we were in murderer's row, in the penitentiary, arrested without warrant, extradicted without warrant, and under the death watch! We had been kidnapped in

> the dead of night and did not know whether our lawyers were aware of our destination. Certainly no one could have expected that we would be put in the penitentiary without a hearing, without a trial, or even the semblance of an investigation.

Actually the legal complaint charging William D. Haywood with murder was sworn to by Owen M. Van Duyn, county attorney of Canyon County on February 12, 1906, and a warrant of arrest for Haywood was filed on February 18th. That same day, an order was issued by the Probate Court of Canyon County stating that the county jail was unfit for the detention of the prisoner, so he had been ordered transferred to the Idaho State Penitentiary. And that was the start of a long, involved court battle that would make headlines that shocked a nation.

CHAPTER 24.

My only hope is to save my soul.
. . . Harry Orchard

The day after the special train rolled into Boise, the fight began. Edmund Richardson, the Federation attorney from Denver, arrived in the city. Three days later, application for writ of habeas corpus for each of the three prisoners was filed in the Idaho Supreme Court. The first shot in the battle of legal warfare had been fired, and it would not be until May, 1907, that the first of the union leaders would be brought to trial.

The same day that Richardson arrived, Steve Adams was arrested at Haines, Oregon, where he had been living under the alias of Steve Dickson and working as a ranch hand. By an odd coincidence, he was arrested by Sheriff Harvey K. Brown of Baker County, Oregon, the same law man who had figured so prominently in identifying Harry Orchard in Caldwell.

By planned coincidence, on February 20th, details of Orchard's confession were made public in part and involved Federation officials in the murder of the former governor as well as other crimes. Governor Gooding, after officially stating that Orchard had made a full confession as to the motive and manner of the Steunenberg murder, told the press the manner in which the confession had been obtained:

> Mr. McParland was aided in his work by Orchard's early training. In his boyhood the Bible was read every night and morning by his parents. The impression of the early days came up and smote his conscience when he was brought face to face with his God.

The mass murderer was quoted as saying, "My only hope is to save my soul." Clarence Darrow, soon to join the legal defense for the Federation officials, had a slightly different opinion concerning the sudden conversion of Harry Orchard. "Where is the minister who converted Orchard? Ha! Ha! A Pinkerton as an Evangel!" he remarked.

Steve Adams was brought to the Idaho Penitentiary, and

against McParland's instructions was not placed in a solitary cell, but was placed with his former partner in crime, Harry Orchard. At first, Adams pretended he did not know Orchard, but later when no one was around they began talking. Orchard refrained from mentioning his own confession at first, but told Adams that since so many people knew of his affairs in crime, he might as well confess to them. He waited for a reaction, but there was none. Orchard had been given no definite instructions from the warden concerning Adams, but it had to come out eventually — the official confession and the implicating of Steve Adams. Apparently that resulted in a fight, but Orchard pleaded with him to make a similar acknowledgement.

The following morning Adams refused to talk and demanded an attorney. What was said between Adams and his lawyer is not known. At any rate, the attorney immediately made a trip to Denver to see the Colorado governor, seeking a promise from him that Adams would not be extradited to Colorado if he aided the State of Idaho in its prosecution of the union leaders. When McParland was informed of what was happening, he was infuriated, feeling that such action placed the prosecution in a bad light.

On February 26, the detective met with Adams in the warden's office, gave him a cigar, told him he was his friend, and related to the killer his own past history. The two had a lengthy discussion on one's duty towards his society and family in general, and McParland repeatedly told Adams that the state always acted fairly with those who acted fairly toward it. They had a good dinner, cigars, and more talk. McParland was priming the prisoner. He went on to mention there was no case against Adams in Idaho, unaware that Adams had been connected with the death of two claim jumpers in the northern part of the state. Adams was in a tight spot and knew it. The thought of hanging was frightening. When the interview was complete, the killer was again put in the same cell with Orchard, who talked to him about the necessity of backing up his own story.

The following day, seven days after being taken into custody, Adams began dictating a formal confession. W.B. Hopkins, McParland's assistant, recorded the material as the murderer blurted it out. Adams was shaking; sweat covered his hands and face. McParland carried a pocketful of cigars, and one by one they disappeared into the killer's mouth — sometime smoked, sometimes chewed. The next day he continued his narration with Mr. Heubner, Warden Whitney's private secretary, furiously taking notes as Hopkins was completing yesterday's work. When Steve Adams was finished with his revealing document, McParland remarked: "His confession is far more important than that obtained from Orchard." The crimes mentioned in Adams' narrative corroborated Orchard's statements on all material features.

On March 1, 1906, the Canyon County Grand Jury began investigating the charges against the accused prisoners in the Steunenberg murder case. Orchard was moved to Caldwell to testify. There Orchard was placed in the Saratoga Hotel. By some trick of fate the only room available was number 19. With no hesitancy on his part, the murderer accepted his quarters. What thoughts passed through his mind that night remained forever locked in Harry Orchard's mind, but the men who took turns guarding him noticed he slept well. The prisoner awakened at an early hour and announced that he felt more tranquil than at any time in the last five years.

At the Masonic Hall, where the proceedings were being held, a large crowd of curiosity seekers had crowded around the proclaimed killer and his guards. Orchard was visibly frightened — the pupils of his eyes were dilated, and the color had drained from his face. As a result of his testimony, the Grand Jury was able to make a report to Judge Frank J. Smith. The judge glanced over the typewritten papers, then handed them to the clerk of the court with the order: "See that bench warrants are issued at once for the men against whom true bills of indictments have been reported and give them to the sheriff to serve at once."

The true bill also bore the name of John L. Simpkins, still at large. One of the last recorded traces of Simpkins was a bank draft drawn by the First National Bank of Denver on the National Bank of Commerce, New York, to the order of W.D. Haywood, secretary and treasurer, and endorsed on the back: "Pay to J.L. Simpkins by Matilda Simpkins (wife)." The draft was cashed at the Traders National Bank, Spokane, Washington, and was dated January 17, 1906. The old dynamiter had vanished so completely, he was never located.

On March 15th, Harry Orchard was taken to Caldwell and arraigned in the district court of Canyon County on an indictment by the Grand Jury charging him with murder. The indictment was read to the prisoner, but since he was not represented by legal council, the court appointed two attorneys to advise him of his statutory rights. When he was asked if he was ready to plead, Orchard answered, "I have no plea to make." Judge Smith directed the clerk of the court to enter a plea of "not guilty" upon the records.

During the reading of the indictment, the murderer stood rigid as a statue, head slightly bowed forward, eyes fixed on the floor. His right hand held his derby hat, and his facial muscles occasionally twitched, betraying his inner emotions. The killer was revealing a little known part of him to the courtroom.

Surrounded by ever-present guards, the murderer was conducted to the depot just before the four-thirty p.m. train was to leave Caldwell for Boise. A large crowd had gathered. As a safety precaution, Orchard was locked in the baggage room. At Nampa, a crowd again assembled to stare at the notorious Mr. Orchard, so the same security measures were again applied. In Boise, the situation was just the reverse — very few spectators. For Orchard, the crowds had been a frightening experience. Mob action was not something new to the West, and Idaho had had its share of such action. McParland believed that Orchard could possibly become a target for an assassin trying to prevent his testifying. The mass slayer also feared the prospect of an unknown killer's trying to silence him before he was brought to trial.

Steve Adams, seeking to prove his confession, agreed to lead investigators in a search for the liquid explosive known as "Hell Fire" that he had buried in Pocatello, Idaho, in 1903.[73] Adams had been sent by Haywood to destroy a carload of strikebreakers on their way from the Coeur d'Alene mining district to Cripple Creek, and he had intended to throw the "Hell Fire" on the railroad car, but fortunately for the strikebreakers, the train on which they were riding had been routed to another line. Adams had carried the liquid explosive in glass bottles in a ten gallon lard can which would fit in his telescope-type suitcase. Not wishing to haul the material around any farther, he had buried it in an abandoned building near the railroad yards.

Late in March, 1906, surrounded by lawmen, Adams returned to locate the buried substance. The killer found the old building with little trouble, but a search revealed nothing but dirt. Adams was positive this was the spot. Just then a man named George McConkey came forward and said that a little over a year ago he was engaged in removing timbers from the deserted mill and had discovered a tin box which contained glass bottles. They had been found in the exact spot Adams pointed out. McConkey had taken the glass stopper out of one of the bottles and in doing so had spilled the liquid on his hands and clothing, burning himself severely. He had thrown the bottle away and had buried the others under a nearby cinder pile.

A few days later, another man working near the old mill had seen blue smoke coming from the rubbish where the first bottle had been thrown. He and another fellow worker had hurried to the spot and began stamping out the smoldering fire with their feet. The liquid had rubbed off on their shoes and had burned the leather to a crisp, confirming Orchard's description of the deadly "Hell Fire" and the purpose for which it had been intended.

Not to be outdone in the "Hell Fire" department, Colorado investigators dug up the ground in the rear of Orchard's house in Independence and uncovered the major ingredients for the liquid substance exactly where the murderer said it had been buried. For safe keeping, the evidence was put in the safe of the local Mine Owners Association with the understanding it was not to be removed until the time of the impending trial.[74]

During March, 1906, another important fragment of evidence confirming Orchard's confession was discovered. W.H. Scholenberg of Wallace was cutting ice in the river near the fire station which was close to the city jail. At about eight o'clock in the morning on the fifteenth, his axe struck a hard object. Deputy Sheriff Hicks was called. He took one look at the frozen bomb and carefully finished digging it out. Then he took it to his office in a gunny sack. He unloaded the bomb and turned it over to Sheriff Sutherland, who took it to the Pinkerton Detective Agency in Denver. The bomb casing was traced to Roach, the plumber, who recognized it as the one he had made for Orchard. The last time the killer had seen the bomb was the day he had given it to Cunningham. Bradley, the proprietor of the Bi-Metallic Hotel, apparently had found out about the explosive device in Cunningham's room, and had taken it from him, and had disposed of the bomb by throwing it in the Coeur d'Alene River.

In May, McParland visited Ida Toney Orchard in Cripple Creek. In spite of the circumstances, she was still extremely loyal to her nomadic husband and claimed that if Harry were convicted, she had enough evidence in a trunk to hang all the officials of the Western Federation of Miners. However, she steadfastly refused to produce the evidence for McParland. Mrs. Orchard maintained that her husband had always treated her nicely and had provided for her needs up to the time he left on his trip to Wyoming.

For some reason the murderer refused to contact his wife, although it would have helped his case had he developed a warmer relationship. One person in the mining district for whom the killer did show genuine affection was the son of John Neville, Charley. Orchard was afraid that if the young man remained in Cripple Creek, he too would be assassinated. Charley had been a witness to his father's attempt to blackmail the Federation leaders and could, no doubt, have introduced damaging testimony concerning a conversation he had heard between his father and Orchard during the Wyoming adventure.

Young Neville was working full time at the Isabella assay office. At the insistence of Orchard and since he might be a good witness for the prosecution at the coming trials, McParland obtained the consent of Mrs. Neville to take the boy out of Cripple Creek to board-and-room with one of the stenographers in McParland's office. The cost of taking care of him was to be paid by Bulkeley Wells.

As the days passed, Harry Orchard, the notorious murderer, was becoming more and more religious. Soon after his arrest, he received a Bible from Dr. David Paulson, of Hinsdale, Illinois, who had learned of the case through newspaper coverage. At first the killer refused the gift, but later he asked for it. After a dark period of brooding, his first use of the Bible seems to have been an attempt to determine whether or not a murderer could be forgiven for his sins. As he read the scriptures, he thought of his days in Canada and the wife and daughter he had deserted.

Finally, on May 12, 1906, he wrote to Mrs. Albert E. Horsley. In the letter, which must have taken a special kind of gall to write, the deserter and bigamist, mentioned that he had taken some pictures of Florence and little Olive with him when he left and was still carrying them with him. He signed the letter with a name that also must have been a bit foreign to him after all these years — Albert E. Horsley. The years of silence were broken when his wife responded, and as he read her letter, the murderer of many men "broke down" and cried. A correspondence was begun. He sold his watch and other personal belongings and sent the money he received to his wife and child. At the same time, Orchard's relationship with his second wife was finished forever. She was an embarrassment to him, a reminder of his life of crime, and he was bent on making a strategic retreat to the womb of righteousness. Meanwhile, the forces of justice prepared a relentless assault that was to gain international attention.

CHAPTER 25.

I know I have been an unnatural monster, but the dear Lord regenerated me, so he could use me.
. . . Harry Orchard

From the moment of the arrest of the Federation leaders, their cases assumed a stature that transcended state and regional interest. The case quickly drew national attention, and before they were finished, they had gained international significance. The charges of kidnapping, of persecution rather than prosecution, echoed and reverberated in the press. Even the conservative press doubted whether the State of Idaho was justified in placing the union officials on trial based on the confession of a man with Orchard's unsavory reputation. The confessional document had astonished millions with its implications.

As each side began building its case, the President of the United States, Theodore Roosevelt, instructed his Attorney General, William H. Moody, to conduct an investigation of the whole affair. Moody assigned an assistant, Charles H. Robb, to do the actual leg work. Robb wrote to James Hawley and asked for a summation of the state's case. Hawley responded by sending a specially prepared report on April 13, 1906. The first paragraph of the report bluntly stated the prosecution's hypothesis:

> The theory of the prosecution in the case against the parties charged with murder is that the "Inner Circle" of the Executive Committee of the Western Federation of Miners planned the affair and procured Orchard to commit the deed.

The report went on to trace the history of the union movement from its conception to its involvement in the various mining states of the West. Also included was an abridgment of the Orchard and Adams' confessions.

Hawley's report impressed the attorney general, who favored prosecution of the defendants and forwarded the document to the president for his study. Roosevelt reviewed the situation and compared the organization of the Western Federation of Miners to the Molly Maguires. Later the president described the three union leaders in a letter to a New York congressman as "undesirable citizens." When the letter by the president was publicized, labor factions were incensed. They accused the President of the United States of prejudicing the case. Radicals took the remark as an indication of a conspiracy existing between government, the mine owners of Idaho, and the Pinkertons. The political party of the opposition took the ill-timed remark as a godsend as they looked toward the next national elections.

Roosevelt countered by saying that their protest would not change his own personal judgment of those leaders who incited people to violence as a means of gaining their objectives. Meanwhile, demonstrations were held in nearly every large city across the country. Union men, on the advice of Socialist Eugene Debs, wore large buttons reading, "I am an Undesirable Citizen," showing the extent of power which the Federation leaders had assumed. By that time, the case had developed into an American crisis as well as one of the most controversial court cases of the twentieth century.

The Steunenberg murder case and the issue of law and order became the major planks in Idaho's election of 1906. President Roosevelt again entered the picture when he publicly endorsed the re-election of Governor Gooding. Perhaps he believed the results of that election would influence the impending trials. In any event, in a personal effort to strengthen the campaign, the President of the United States sent his secretary of war, William Howard Taft, to Idaho. Gooding's opponent was Charles O. Stockslager, the chief justice of the Idaho Supreme Court, the man who helped bring the Pinkerton Agency into the murder case and had ruled on the habeas corpus writ. Governor Gooding was re-elected in a close election, and the Republican party swept all state offices and both houses of the legislature.

While this heated debate raged across the country, the man who started all the furor languished in prison, comforted only by the daily reading of his worn Bible. Harry Orchard walked, talked, and slept in his own special hell while the country chose sides in the case.

In the spring of 1906, Bulkeley Wells and Deputy Warden Mills of the Idaho Penitentiary, took Steve Adams to Telluride, Colorado, in an attempt to locate the body of J.W. Barney, a non-union shift boss of the Smuggler Mine, who in 1901, had been murdered by Federation hoodlums. Adams claimed he helped bury Barney. The search was a failure. At any rate, Adams' cooperation with the prosecution was to come to an end shortly, but until then he lived with his wife in a cottage at the prison and smoked cigars furnished by the state.

When Adams took exercise around the prison yard, he occasionally went up on the gallows platform and took a nap. He took great delight in his antics. Orchard, on the other hand, would not go near the structure. Adams was also a very frightened man, unable to shake the fear that he too would be assassinated by Federation killers. The defense recognized only too well his worth and importance to the prosecution. He was a crucial figure in their war plans because Idaho state law placed him in such a position:

> The statute of the state imposes a bar to conviction on the testimony of an accomplice alone, no matter though he may be believed by the jury, unless there is other independent evidence tending to connect the defendant on trial with the commission of crimes.

Adams represented the "other independent evidence," but unlike Orchard, the defense found Adams could be reached, convinced and led to repudiate his confession.

One of the few persons who had access to the killer was James W. Lillard, of Baker County, Oregon, an uncle to the girl Adams had married. Darrow saw Lillard and skillfully convinced him that his nephew had made a grave error by confessing and turning state's evidence. He suggested that if Adams would repudiate his confession, and if the state in return would bring him to trial, Darrow himself would act as his defense counsel.[75] It was a promise of magnitude, but without Adams, the odds were that Orchard's testimony alone would never convict the union officials.

Lillard became sold on the plan, and during a series of conversations with his niece, he convinced her that if her husband admitted he had made a mistake in his story, he might have a chance to escape death or a lifetime in prison. Letters from his brother Joe and Mary Ann Mahoney also persuaded him to repudiate his confession until he finally agreed and was released on a writ of habeas corpus in the district court of Ada County on September 8, 1906. The State of Idaho made no effort to restrain the murderer, and he was released from custody. The same day, however, Shoshone County authorities arrested him and charged him with the death of the two claim-jumpers on Jack Simpkins' property in the northern part of the state in 1904.[76]

After a preliminary hearing in Wallace, the prisoner was bound over for trial for the murder of Fred Tyler, one of the two dead claim-jumpers. The trial was set for January, 1907. By this time the United States Supreme Court had removed the last legal roadblock to the trial of Haywood and his two accomplices. In this respect much was at stake in the Adams' trial — the state needed his testimony in the impending trials of the labor leaders. If Adams could be convicted of the murder of Fred Tyler, that would give the prosecution all the leverage it needed to force Adams to turn state's evidence once again. Adams' confession included a description of the killings:[77]

> We went in one night, Newt Glober, Albert Mason, and myself, to Simpkins' cabin, made some coffee and had breakfast, and in the morning we went over to try to catch Tyler at his cabin on Jack Simpkins' claim. He was gone, so we laid there until sundown . . . I saw it was Tyler coming. He had a big gun buckled on him. I stepped out on the trail and told him to throw up his hands. We disarmed him, took him to Simpkins' cabin, stopped there until next morning, had breakfast there, took him three miles out in the timber next morning, and I killed him.

The killing of Boule was similar.[78] Adams met Simpkins at Harrison right after the Tyler killing and told him what had happened:

> . . . In about a week or thereabouts, or probably two weeks, we went back up in there and met a fellow named Boule, and another man coming down the trail near Simpkins' cabin. We opened fire on them, killing Boule — Jack Simpkins, and I, and Newt Glover . . .

On the basis of this admission, Adams was held for trial. Orchard added some details that Adams and Simpkins had related to him until the actual sequence of events was pretty well laid out.

As the Adams case was being prepared, McParland had some misgivings about it, and he wrote in October, 1906:

> In connection with the matter, while I think an acquittal in the Adams case impossible, still we might have a hung jury. That of itself would give him courage. At the same time the defense would be in possession of a good deal of my testimony to help them at Caldwell.

McParland's words would prove to be only too true.

The Adams trial was to be a preliminary bout for the main event coming up. James Hawley represented the state, while the work of the defense was handled by Darrow and Richardson. The trial was highlighted by the big-time, colorful opponents — plenty of footwork, punches and counterpunches, and an abundance of legal strategy. When a bundle of moldy bones, claimed by the state to be that of the deceased, Fred Tyler, was introduced as evidence, the court gasped. The mother of Tyler could hardly contain herself as she identified the disintegrated clothing that had covered the bones as that once worn by her son.

The defense went to lengths to show that the defendants had an alibi established by friends. Then Adams, with the look and tone of a wronged man, testified that his confession had been extorted from him by threats of death and extradition to Colorado to stand trial. In the end, the giants of the legal world had battled to a draw. The only thing that was "hung" was the jury, as they deadlocked six-six. Round one was over. If Adams had been acquitted, the state was prepared to arrest him immediately on the charge of murdering Boule.

Steve Adams was tried again at the conclusion of the Haywood trial in the fall of 1907. The state still gambled on the outside chance that they could get a conviction and force the killer to testify in the coming trial of Pettibone. It was a replay — Hawley for the prosecution and Darrow for the defense — resulting in another deadlocked jury. Steve Adams passed forever from Harry Orchard's life; he had wagered everything on the Darrow offer and had won. He was never again brought to trial in Idaho.

A most unusual letter was written to Orchard in April, 1906. Because of its nature and the emotional condition of the prisoner at that time, it was not delivered for a full year. The writer was an old friend from the Coeur d'Alene days, Dan Gainey. Completely unknown to the mass slayer, Gainey had been standing on the platform of the Independence depot when it was blown up and had been permanently disabled. Gainey wrote:

> . . . You perhaps know by this time I was on the Independence Depot platform when it was blown up. I lost one foot and had the other so badly broken that it can never be well again. For nearly two years I have suffered undescribable mental and physical tortures . . . And now, Harry, I cannot believe you were always cruel and depraved, but rather that you were persuaded and led into these crimes by that specious reasoning of the Miners' Union which declares the end justifies the means . . . .It is altogether probable that providence is using you as an instrument of divine retribution against that organization which is responsible for your inhuman conduct and moral degradation . . .

When Orchard read the letter he broke down and wept. He clutched the letter, reread it, and turned to his Bible for solace and for answers. After that, an intense correspondence began between the murderer and his victim. Orchard responded to Gainey, "I know I have been an unnatural monster, but the dear Lord regenerated me, so he could use me. I have told the whole truth; I did not shield myself or anybody else, nor did I wrongfully accuse any innocent person." Orchard was to hear more about that fateful event. Phil Chandler, a guard at the prison during this time, had also been standing on that ill-fated platform and had suffered a broken leg.

As the months in solitary confinement passed, the murderer became fully immersed in the spirit of religion. Through the efforts of McParland, the brooding Orchard met Edwin S. Hinks, dean of St. Michael's Cathedral in Boise. After a short introduction and conversation, the killer asked an opinion from Mr. Hinks, and the answer was to be based strictly on holy scripture, "Was he, as a murderer, shut out from the hope of God's forgiveness?" Orchard was obsessed by two concerns — one was a concern for his immediate future, and the second concern was for his future after death. These questions gnawed at his guts and his mind. He became desperate for an answer. Hinks told him that neither in the scriptures of the Old Testament nor the New Testament was there a single word to preclude an honest penitent to God, who offered forgiveness and pardon. His answer gave the murderer hope and purpose; it was a soul-satisfying response to an inquiry by a tortured, lonely man who wanted to believe in a God who could forgive.

Hinks also gave a most interesting observation into the nature of the mass slayer:

> As I comprehend the transformation of Harry Orchard from reckless criminality to a penitent willing to tell the truth, I feel that the world should understand that his change of front was not in the order of religious conversion, then moral perception leading to confession. No! It seems to me the order was first physical, second moral, and finally religious.

Finally, it was announced that the case of William D. Haywood would be heard the month of May, 1907. A change of venue had been granted, and the trial was scheduled to be held in Boise, Ada County. Both the prosecution and the defense sought to strengthen their cases, hiring private detectives to marshal evidence and secure other material. They were all aware of the importance of each minute point that might be a part of what would become known as the Haywood case. For some it would be the moment of a lifetime. History and time would evaluate their performances.

Harry Orchard, in the meantime, had been writing his legal wife regularly and was concerned about her well being. He was aware particularly of the possible embarrassment to her and the child when the Haywood trial began. Mrs. Horsley was working as a housekeeper and was barely making ends meet. Governor Gooding, to pacify his star witness, informed Orchard that she would be taken care of financially for the time being. Nevertheless, Orchard continued to worry about her; the thought of her being subjected to violence or revenge began to weigh heavily on his mind.

McParland realized a move was necessary, so he arranged to have Mrs. Horsley sell her furniture and secretly be moved to the Denver area. Upon arriving in the capital city, she and the child were to report to Mr. Cary, a Pinkerton employee, and she was to use her maiden name of Florence Fraser. From now on when she wrote to her husband, she was instructed not to date her letters or designate any particular place, but simply to write them and give them to Mr. Cary who would forward them to Orchard. From that point on McParland did not want to know anything about the situation. He explained his reasoning bluntly in a letter to Orchard.

> I have not seen your wife since I arrived here, and I think it best that I do not see her so if I am used as a witness and the question comes up to me as to whether I have met her or know where she is presently located, I can be in a position to reply that I have not met her, and don't know for a fact where she is located. However, you can advise her to call on Mr. Cary or in his absence on Mr. Pettyman anytime she wishes. If she is in need of funds, you simply tell the warden, and he will communicate with the Governor and then the Governor can communicate with Mr. Cary.

As a result of McParland's efforts, Mrs. Horsley spent the critical time during the trials in Denver under the protection of the Pinkerton Agency, living quietly under her maiden name.

Pacified once more, the dynamiter fell back into his routine of Bible reading and answering the multitude of questions put to him by the prosecution and its investigative staff. The killer was also working on the story of his life, which proved to be an exercise in soul-searching. And as he wrote, the trial of Bill Haywood was fast approaching.

CHAPTER 26.

Gentlemen of the jury, the great question in this case is, we admit it, has Harry Orchard told the truth or any part of the truth?
. . . William E. Borah

How can anyone believe a self-confessed perjurer, kidnaper, thief, firebug, and multiple-murderer?
. . . Clarence Darrow

The most extraordinary man I ever encountered in probably as long a life of criminal prosecution as any man in the state was ever called upon to engage in.
. . . James H. Hawley

When the trial of William Dudley Haywood, secretary-treasurer of the Western Federation of Miners, commenced on May 9, 1907, it was immediately evident that this would be no ordinary murder trial. It was to be a trial of such magnitude that it somehow seemed out of place in the quiet, tree-lined but obscure city of Boise, Idaho. There was the ever-present feeling of excitement and tension, but there was no outward expression of fear or violence. Regardless of what happened in the courtroom, the city and its population took it in stride; famous names and persons seemed to disturb its tranquility very little. The State of Idaho had pledged that Bill Haywood and the other defendants would be given a fair and honest trial, and it was doing its best to live up to that vow.

Fremont Wood was designated as the presiding judge. Wood had been serving as the United States attorney for Idaho during the 1892 Coeur d'Alene mining troubles, and then he served as the prosecutor of the federal cases that grew out of the incident. At the same time, the opposing attorney for the accused miners had been James H. Hawley, and among his clients had been the former justice of the peace, George Pettibone.

The chief prosecutor for the State of Idaho, James H. Hawley, has been recognized by historians as the most outstanding trial lawyer in the history of the state. A former miner himself, Hawley had been a member of the Territorial Legislature, mayor of

Boise, and a power in the Democratic party. With his good friend William E. Borah, he served as special state prosecutor in the Paul Corcoran murder trial, a case that stemmed from the 1899 Coeur d'Alene mining disturbances.

William E. Borah, state prosecuting attorney, assisted Hawley. Borah had been a part of the case from the moment the special train, commanded by Governor Gooding, had left Boise and headed west for Caldwell and the scene of the Steunenberg murder. Borah himself had just been elected United States Senator.

The staff members of the defense were also recognized for their considerable legal ability. Edmund F. Richardson had represented the Western Federation of Miners in long and costly wars and was a particularly skilled cross-examiner in the courtroom. Clarence Darrow, the labor lawyer from Chicago, was the man who had defended Eugene Debs in the Pullman strike case and was on the threshold of a national acclaim. John Nugent, a Silver City, Idaho, attorney and a personal friend of Bill Haywood, had entered the case at Haywood's request. Edgar Wilson of Boise had been a member of the United States House of Representatives and a former law partner of Judge Wood. Peter Breen of Butte, Montana, had been an active organizer and dynamiter in the North Idaho mining war of 1892, and in 1899 he had been a member of Corcoran's defense staff. Fred Miller, the Federation attorney from Spokane and Harry Orchard's lawyer for a time made the team, as did Leon Whitsell, another minor figure of the defense.

On May 16, during the impaneling of the jury, Harry Orchard, the chief witness against the Federation leaders, finally made a public appearance. For the occasion, a group of newspaper correspondents were admitted into the inner confines of the penitentiary to see for themselves the mysterious and controversial self-confessed murderer. Ward Whitney told them that no questions relating to any features of the case were to be asked. Proper questions would be allowed, although the prisoner might not be willing to answer them all.

Seats for the visitors had been carefully arranged in the warden's office. As the reporters walked into the room, a man wearing a gray suit rose from his chair at the opposite end of the chamber. "Gentlemen," announced the warden, "this is Harry Orchard." Orchard nodded to the visitors. He was perfectly at ease, and a thin smile played on his lips as he spoke. "I'm very pleased to see you," he said. This was the reporters introduction to the man who claimed he had killed more than a dozen men in one wild moment of madness. The interviewers were stunned. This man they saw was intelligent and polite; he had twinkling blue eyes and a well-kept brown moustache. Short, somewhat stoutly built, he carried himself well. Above all, he was most pleasant to speak to. One reporter remarked: "There was about him an air of composure and serenity which expressed more plainly than any words he could utter the fact that he had reached a thoroughly satisfactory solution of his problem."

A. E. Thomas, representing the *New York Sun*, probably summed up the group's feelings as clearly as any one man could:

> Whatever the Harry Orchard of December 30, 1905, may have been, men who look like the Harry Orchard of May 16, 1906, are men who tell the truth. There is a conscience behind those blue, unfaltering eyes of his. How it came to be there, how it germinated, how it grew, is a story that will sometime be told. But that time is not yet.

The newsmen may have been impressed with the whole event, but not Judge Wood. When he read reports of the visit, rage burned inside him. He was particularly disturbed by a Boise newspaper account which concluded that Orchard was telling the truth. It had been the custom of the judge to request jurors not to read the papers' running account of the trial, but he had no control until a prospective juror had been drawn from a general panel.

At this particular time there was a panel of jurors ready to be interviewed. His Honor had the courtroom cleared of all jurors and prospective jurors, then referred the matter of the reported interview to the Ada County prosecuting attorney, Charles F. Koelsch, and instructed him to make a thorough investigation of the circumstances and report the facts and findings together with his recommendation to the court. Koelsch did as he was requested, found no intentional violation amounting to comtempt on the part of the reporters involved. And so the incident was carried no further.

With the jury selection completed, the actual Haywood trial commenced on June 4. The state began to build its case. Witnesses were called to prove that the victim, Frank Steunenberg, was indeed dead, murdered in Canyon County, Idaho. Next, Jack Simpkins, a member of the executive committee of the Western Federation of Miners, was positively identified as Harry Orchard's companion during the time the state claimed the murder was being planned. The prosecution was planning to lead up to the introduction of the alleged murderer himself and to his direct testimony in the case.

June 5th proved to be the high point of the Haywood trial, the day curious thousands had been looking forward to. Finally, breaking the brief silence was the deep-throated voice of Hawley as he intoned, "Call Harry Orchard." The door opened, and a guard entered. The courtroom spectators were tense. Behind the guard stood Harry Orchard, a short, stout, well-groomed man. His eyes were fixed on the floor as he walked to the witness stand.

Haywood's god eye focused on the killer, a look of hatred there, intense as the heat of hell. Arriving at the elevated witness stand, Orchard stood with his right hand raised as the clerk of the court administered the oath. Then he looked at the other accused man. At last, it was Harry Orchard's turn to speak to the world, to tell his story. Gradually, Orchard gained control over his emotions and quietly responded to the questions. As Hawley led him through his testimony by going from event to event in chronological order, it was evident to the courtroom that Orchard had been well rehearsed for this appearance.

At three o'clock in the afternoon, Hawley complained to the trial judge that he had a severe headache and asked for an adjournment. Orchard, too, was exhausted by the ordeal. It had been mentally fatiguing to all. Once he had overcome that first question and the impact of the situation, Orchard had answered question after question by narrating the brutal details of his participation in crime. His manner was easy and confident, and at no time did he seem to be bragging or belligerent. He was what trial attorneys like to call "the perfect witness." His only fault was that he talked too low to be understood any distance from the stand, and several times he was admonished to speak louder.

When Hawley had finished with the witness, he was turned over to the defense for cross-examination. Richardson, whose reputation had been made by his ability in this particular area, took charge. The cross-examination of Orchard lasted twenty-four hours and was spread over a week. It soon became apparent that the murderer had no qualms about answering any of the defense counsel's questions, even if it involved probing into his past life in Canada. Unhesitatingly, he admitted to crimes not previously covered in the testimony. His memory was phenomenal. Richardson was unable to shake Orchard's testimony. His sharp, penetrating questions did not throw the witness. Even Darrow was impressed with the performance.

At the conclusion of the presentation of the state's case against Haywood, the defense rose and "requested the court to give to the jury an advisory instruction for a verdict of not guilty on the grounds that under the statute there was no corroborating evidence of Orchard's testimony to justify a conviction." It was a moment of truth for Judge Wood; the trial and case against the Federation leaders hung by a very thin and worn thread.

Wood had always maintained that there was very little corroboration of evidence at this point to warrant submitting the case to the jury. Perhaps if his former law partner, Edgar Wilson, had not been a member of the defense staff, the judge might have decided in their favor, but under the circumstances it might appear as if he were deliberately throwing the case because of their previous relationship. Wood chose to uphold the honor of the court when he stated, "Gentlemen, the court is clearly satisfied that his case would be heard and a jury would decide the fate of the accused."

Years later, Judge Wood elaborated on the incident saying:

> The principal portion of the corroboration in the Haywood case was developed through the witnesses of the defendants when making their defense, and while there was no corroboration against Haywood in the original

presentation by the State, it was my opinion that the corroboration was ample when the case against Haywood was submitted . . .

Clarence Darrow in his inimitable style presented the jury several theories concerning Orchard's testimony. His main contention was the Harry Orchard killed Frank Steunenberg to experience personal revenge for having been driven out of the Coeur d'Alenes and for having to sell his interest in the Hercules mine. Furthermore, he did not blow up the Bunker Hill and Sullivan concentrator as he claimed, for at the moment that he huge structure was being destroyed, Harry Orchard was playing cards with some old cronies in the town of Mullan. Darrow emphatically claimed that the State of Idaho was really prosecuting the Western Federation of Miners rather than the defendant, William D. Haywood. His opening statement to the jury clearly expressed this stand: "Every illegitimate child born west of the Mississippi has been wrapped in its swaddling clothes, hurried to Denver, and laid on the doorstep of the Western Federation of Miners."

Orchard had related how he had planted the bomb in the shaft of the Vindicator mine and arranged it so that when the safety bar was raised, a revolver would discharge into a box of explosives and thus kill McCormick and Beck. The remains of a broken and twisted revolver were produced by the prosecution to corroborate the witness's story. The defense countered with a theory that the two men were carrying a box of powder with them when they went to the ill-fated sixth level. Beck also had had a revolver that he was carrying in his rear pocket, and it apparently fell, discharging on impact with the ground and subsequently detonating the dynamite. The cager at the Vindicator mine during the incident told of seeing two men in the mine who shot at him, to some extent strengthening Orchard's story of his underground experiences. Orchard stated that he had been given money by both Moyer and Haywood for his work. Both men, however, took the witness stand and under court oath denied any knowledge of the crime or having given Orchard money to commit it.

When Harry Orchard related the details of the fatal shooting of detective Lyte Gregory in Denver, he told how he was accompanied by Steve Adams and George Pettibone, the latter man fingering the victim and contributing the murder weapon. Neither Adams nor Pettibone was placed on the stand to contradict the testimony. Instead, the defense contended that Gregory probably had been killed by a personal enemy since the Western Federation of Miners had no grievances against him.

As he went on with his testimony, one of the most dramatic events related by Orchard included the tragic details of the destruction of the Independence depot and the killing of thirteen innocent miners by himself and his accomplice, Steve Adams. The state showed that both Orchard and Adams left the district immediately after the disaster and assumed different aliases. Adams became known as Steve Dickson and Orchard, Thomas Hogan. The defense did not deny that Orchard had committed the slaughter, but it implied that he had been acting as the agent of the Mine Owners Association at the time. Orchard's statement implicating Adams as an accomplice in the mass murder was left unchallenged.

When Orchard told of being sent to San Francisco to kill the old Federation adversary, Frederick W. Bradley, the prosecution and defense deadlocked. Neither side made a clear presentation concerning what actually happened to Bradley when he opened the door to his apartment house. Was it a gas explosion as the defense claimed, or was it Harry Orchard's dynamite that had exploded? There was no official cut and dried solution to that incident. The killer's landlady in San Francisco testified to having found the lead and wood shavings in the room, the screw-eye in the door, and the heavy suitcase belonging to the witness. The San Francisco city chemist was called to testify as to the presence of strychnine that he had found in the Bradley milk. Postal employees testified regarding the registered letter that was mailed by Pettibone's clerk in Denver to Orchard in San Francisco.

Orchard went on to describe his other futile attempts to assassinate, presenting an impressive list of prospective victims to the jury: Governor Peabody, Justices Gabbert and Goddard, and the Adjutant General, Sherman Bell. The state pointed out to the court that the witness had been openly associated with

Federation officials and that he had lived with Pettibone for several weeks. But excluding the killer's testimony, nothing was shown to prove that the accused union leaders had had knowledge of the crafty dynamiter's plans.

The mysterious circumstances concerning Walley's death in the deserted vacant lot, and the details of the planting of the box of explosives in front of Judge Goddard's gate, were examined. The Peabody bomb was exhibited in court, and the Denver plumber testified and identified the bomb casing as the one he had made for Harry Orchard during the time he claimed he had spent tracking down Peabody.

Governor Peabody took the stand and swore he once saw Orchard in Canon City. Vaughan, the insurance salesman and companion to Orchard, told of the unusually heavy suitcase the killer had in Canon City and of the strange, ticking noise inside, adding that he had been shown the clock mechanism and had been told it could cause mine explosions. Humor unexpectedly found its place in the courtroom when the State produced as evidence letters of recommendation that had been written for Orchard at the time he applied to sell insurance. The reading of the letter from Horace Hawkins, law partner of the defense council, Richardson, caused the lawyer to smile with embarrassment.

During the discussion of the bomb buried under Goddard's gate, the defense jumped at the chance to point up the improbability of an explosive device remaining undetected for months under a well-manicured lawn. The inference by the defense was that it had been placed by unscrupulous detectives to be discovered later for corroboration of Harry Orchard's well-publicized confession.

When he testified about the murder of Frank Steunenberg, Orchard was adamant on the stand when he swore that Haywood, Moyer, and Pettibone had all been present when he left Denver for Idaho, and that all three had been well aware of his mission in that state. Haywood gave him money, and an enthusiastic Pettibone had helped him pack his bomb in the trunk and then had accompanied him to the railroad station and there bade him success. The record showed that the assassin had registered at the local Nampa hotel as "Thomas Hogan."

He then detailed the frustrated stalking of the elusive victim, the trip to North Idaho, and his reunion with Jack Simpkins, a member of the executive board of the Western Federation of Miners. Returning to Caldwell, Simpkins had registered at the hotel as "J. Simmons." At this point, the prosecution produced a witness from Wardner, Idaho, C. H. Wentz, who testified that he taught Simpkins how to write and that the signature of "Simmons" on the hotel register was actually that of Jack Simpkins.

After Orchard narrated the details of the murder and of his subsequent arrest, the telegram from Miller was placed before the court as evidence. Miller had wired that he would go to Caldwell to defend the accused slayer. It was shown further that Orchard had made no request for legal counsel and that the Federation had paid fifteen hundred dollars to Miller in Denver to defend him. The visit of the labor attorney, Sullivan, from Denver, was also brought up. It was also shown that Haywood had contacted the secretary of the Federation at Silver City and requested that he employ John F. Nugent to look after the Federation interests following the arrest of Orchard. Nugent had subsequently replied that he could not see that the interests of the union were involved. At the time this had taken place, the only evidence that connected the accused slayer with the Federation was the risque souvenir post card which was addressed to Charles H. Moyer and which was found in Orchard's possession.

Darrow had no misgivings about the fact that Orchard had killed Frank Steunenberg, and that he had been assisted by Simpkins. The murder motive was something else. The Chicago attorney considered it an act of personal revenge in retaliation for Steunenberg's actions during the labor troubles of 1899 which resulted in Orchard's fleeing from the mining district and Simpkins' being arrested and serving time in the notorious Bull-Pen. Orchard was pictured as a man who had been cheated out of a fortune when he was supposedly forced to sell his interest in the Hercules for a mere pittance of its actual value, in order to escape capture.

The prosecution assaulted Darrow's personal revenge motive with solid legal evidence. Deeds and records were produced to show that Harry Orchard had sold his interest in the Hercules mine more than twelve months before he was forced to leave the mining district. When the killer had left, he had no holding or interest in any land, business, or industry.

The defense further asserted that while Orchard was living in Cripple Creek, he had been a detective for the mine owners. The killer had admitted freely that he had received money from D. C. Scott, detective for the Florence and Cripple Creek Railroad. He also had admitted that he informed the detective of a proposed attempt to wreck a train carrying non-union laborers. It was also brought out in court that he had received money and a train ticket to Denver from Scott, so that he could become acquainted with the Federation leaders.

One thing seemed certain. If Orchard had been an undercover agent for the Mine Owners Association, he was a damn poor one, since he had no records of his employers, such as letters or telegrams that he was known to have received in the course of his activities. If he had worked for McParland, he would have been fired in short order for incompetency.

And so it continued. Each verbal clash between the defense and the prosecution was faithfully reported by the national press. Snarls, cutting remarks, dramatic gestures — all were pounced upon and given special meaning as they were interpreted for legions of readers, who eagerly waited for the next development. And they were not disappointed!

CHAPTER 27.

A man of some confidence in himself, at times, and yet timid under other circumstances.
. . . Dr. Hugo Munsterberg

The best was yet to come! As the trial ground on, the fifteenth annual convention of the Western Federation of Miners had assembled in Denver's Odd Fellows Hall. The two hundred or so delegates sent a telegram of morale-building support to Moyer, Haywood, and Pettibone, which stated, "Confident of your innocence and are with you until your honor is vindicated and your freedom assured." A number of delegates were wearing buttons bearing the words, "I am one of the undesirable citizens."[79]

A visit to Boise at this time was not complete without a trip to observe the famous trial proceedings. And many famous personalities turned up in the courtroom. One of these was one of the foremost psychologists in America, Dr. Hugo Munsterberg, professor of psychology at Harvard University. A very special interest of Munsterberg's was the inner workings of the criminal mind. He found a willing subject in Harry Orchard. The psychologist not only observed the murderer in regular court sessions, but also spent eight hours testing him at the penitentiary. Munsterberg claimed Orchard was the most remarkable criminal he had ever examined; he was fascinated with the man:

> Orchard has an alert mind; he is singularly keen. My measurements, as well as my oral examinations of him, show that his mind is not only active and accurate, but he has a very quick perception; he anticipates. In fact, for mental alertness, few Harvard students would measure up to him.

Dr. Munsterberg also made another comment that would seem to be most meaningful in analyzing the killer's mental makeup. "Orchard's emotional life is simply dead," he stated. No doubt many less well-educated persons who knew the dynamiter had already made this same observation. The doctor also pointed out:

> Even if his religious conversion is genuine, his so-called religion lacks also every sound and deep feeling; it is thoroughly utilitarian; he serves God because He will reward him after death.

His examinations showed the subject to be a very complex individual, "A man of some confidence in himself, at times, and yet timid under other circumstances." But perhaps the summation of the psychologist's research was contained in one remark: "Orchard's confession is every word of it true." Since the science of psychology was then in its infancy, the remark had little impact or importance. Thus, it was not used to advantage by the prosecution.

As a defendant on trial for his life, Bill Haywood played his role equally as well as did Orchard. He gave the impression that he was sincere and sure of himself and his facts. As he sat in the courtroom day after day, he was by all appearances a quiet, friendly man. He comforted his aged, invalid mother, affectionately patted his wife's hand, played with and teased his young daughter. In his cell at night he was a different man. The relationship between himself and Moyer had deteriorated into a barely tolerable situation. Moyer had grown sullen during his confinement, and when Haywood uttered a testy tirade against mineowners and capitalists, he would silently turn his head away and shut out the big man's rumblings.

On the day Governor Peabody arrived from Colorado to be a witness for the prosecution, Orchard was having lunch in Hawley's office.[80] When Peabody walked in, the killer almost fell apart emotionally and began babbling like an idiot. Orchard pulled himself together and seemed genuinely pleased to see his once marked victim was still alive. Meanwhile, the prosecution stated its case.

On a hot summer day, Friday, July 19, about thirty minutes after two in the afternoon, James Hawley opened the State's summation. He reviewed for the jury the violence perpetrated by the Western Federation of Miners and pointed out the characteristics of a criminal conspiracy. In his homey, down-to-earth manner, he discussed Harry Orchard and the truth of his testimony.

Richardson was next. The genial attorney launched an attack on the evidence supporting the State's case, making a point of the fact that the well-known opponents of the Western Federation of Miners, namely Peabody, Bell, Goddard, and Gabbert were alive and breathing, proof enough that they were not marked for death.

Clarence Darrow slowly walked up and down before the jury members, stopping only to look the twelve men straight in the eyes. He expounded on the fact that Orchard was a notorious liar, motivated by his own greed, and that the trial itself was nothing more than a capitalist plot to kill the defendant Haywood and involve the Western Federation of Miners.

Senator William E. Borah gave the State's closing argument. He claimed that no man could keep a story as complicated and involved as Orchard's straight, unless it were the truth, particularly when every detail was subject to repeated cross examination and still the story had not collapsed. He dwelt on all the documentary evidence in the case supporting Orchard and the State's case and emphasized that the Western Federation of Miners was not on trial and that Haywood should not be convicted because of his reputation in Colorado.

While Borah was discussing Haywood and the accusations against him, the defendant's mother collapsed. Reporters interpreted this as an indication of her belief that Borah's accusation of her son was the truth. It proved a final thrill to the Haywood trial.

On July 27, when the summations were completed, it was the trial judge's turn to address the jury. Little did Judge Wood realize that his remarks to the twelve men that July day would permanently injure his personal reputation and his career. He presented sixty-five instructions to the jury. One which had a great deal of effect on the trial dealt with the question of reasonable doubt:

> The burden is on the prosecution to prove beyond a reasonable doubt that a combination and conspiracy was

formed, and that in the execution and carrying out of such a conspiracy and design, some of the parties to said conspiracy and design, some of the parties to said conspiracy and agreement killed Frank Steunenberg. The burden of establishing these facts is upon the prosecution throughout and never shifts to the defendant, and, therefore, if the prosecution has failed to prove beyond a reasonable doubt each and every one of these facts, you should acquit him. And you are further instructed that such a combination and conspiracy cannot be established by the uncorroberated testimony of an accomplice.

Judge Wood particularly emphasized the need to corroborate Orchard's testimony and confession by witnesses who were not parties to the alleged conspiracy.

By his own admission, Judge Wood believed that Orchard's testimony was true, and he fully expected a guilty verdict to be brought in against Haywood. Why then the extensive jury instrucitons? The most logical explanation could be the honest desire of the judge to make certain the defendants had a fair trial, one that could stand up to a review by a higher court.

The fate of William D. Haywood now rested with twelve men as they retired to deliberate. It was to be an all-night session, a trying ordeal. The following day, Sunday morning at seven-thirty, court was reconvened to hear the decision. The foreman handed the judge the verdict in a sealed envelope. The room was silent. The defendant seemed relaxed and the atmosphere expectant. The envelope was slowly opened, and Judge Wood read the decision silently, then passed it to the clerk of the court, who read the statement:[81]

We, the jury in the above entitled cause, find the defendant, William D. Haywood, not guilty.
Thomas B. Gess, Foreman

It was over. Bill Haywood was a free man. Exuberant, bursting with confidence, he was his old self again. The acquitted man returned to his cell to pick up his personal effects.

The verdict of the jury was met with mixed reaction. Most of the unprejudiced court observers seemed to agree that the prosecution had failed to prove its case. In the minds of many, the verdict would have been more appropriate had it been "guilt not proven." Big Bill Haywood had not been convicted, but some of the violent methods of the Western Federation of Miners had been exposed. The relentless probing of the court had exposed the abuse and the irresponsibility of leadership. *The Nation,* a leading magazine of that period, made the point clear when it said, "The fact that a labor union had for years made of itself a sort of terrorizing mafia in five states, using guns and dynamite for argument and murder for persuasion, has now been established up to the hilt."

Without Steve Adams' testimony, the prosecution lost most of its important corroborating evidence. At best one can only speculate what the results would have been had Adams testified for the State. McParland did not consider him the equal of Orchard as a witness for he lacked the fine ability to fix dates, names, places — the ability that assisted Orchard so well. Perhaps Borah had been right when he had urged that the state concern itself only with the prosecution of Harry Orchard for the murder of Frank Steunenberg and let the State of Colorado deal with the leaders of the Federation. Whatever the case, the State of Idaho failed to prove that a conspiracy had existed. Nobody felt more bitter over the jury's decision than Harry Orchard. Most of his anger was directed at Judge Wood who, he claimed, "emphasized reasonable doubt so often the jury evidently thought he had a reasonable doubt."

Rumors, accusations, and innuendoes were rampant after the verdict was brought in. Judge Wood was accused of throwing the case. Darrow was accused of bribing members of the jury. Borah was accused of having allowed his political ambitions to influence his courtroom work. The Western Federation of Miners was accused of packing the courtroom daily with such a show of force as to intimidate the jurors. Even the bailiff did not escape accusation when he was accused of taking a bribe to permit access to the jury.

As soon as it could, the state proceeded with Steve Adams' second trial which ended as it had the first time — in a hung jury. This was a last ditch attempt to gain a favorable verdict to force

Adams to be a witness for the state in the projected trials of Pettibone and Moyer.

The trial of George Pettibone sounded like a replay of the Haywood trial with little new evidence introduced. Orchard had his day in court and testified just as forcefully as he had before, but by then public interest had dimmed. The jury rendered the verdict, "not guilty," for Pettibone on January 5, 1908.[82] The case against Moyer was then discharged for lack of corroborating evidence, and Simpkins was never brought to trial. And so the conclusion of the Pettibone trial left the State of Idaho with only Harry Orchard to account for since his accomplice Jack Simpkins was not in custody.

On March 10, 1908, Orchard was brought before Judge Wood and he quietly told the court that he wanted to withdraw his former plea of not guilty and enter a plea of guilty, as charged in the indictment. Judge Wood asked the defendant if he understood that such a move would result in the court having to pronounce the death penalty. If Orchard understood the magnitude of the guilty plea, and by all indications he did, either the mass slayer wanted death or it was a good grandstand play for publicity. Orchard had contemplated suicide before the Haywood trial had every taken place, but there is no denying he enjoyed his part in the drama that had unfolded before the astonished public.

Orchard was advised to consider carefully the matter of changing his plea. On March 18, 1908, Harry Orchard stood before the court awaiting judgment. Before pronouncing the sentence, the presiding judge reviewed the case and concluded among many things that:

> It was the particular province of the Court to observe and follow the witness during the former trials, and I am of the opinion that no man living could conceive the stories of crime told by him and maintain himself under the merciless fire of cross-examination by leading attorneys of the country, unless he was testifying to facts and circumstances which had an actual existence in his own experience. A mere child may testify truthfully and maintain himself during cross-examination. A man of mature years may be able to frame his story and testify falsely to a brief

> statement of facts involving a single transaction and maintain himself on cross-examination. But I cannot conceive of a case in which even the greatest intellect can conceive a story of crime covering years of duration, with constantly shifting scenes and changing characters, and maintain that story with circumstantial detail as to times, places, persons, and particular circumstances, and under as merciless a cross-examination as was ever given a witness in an American court, unless the witness thus testifying was speaking truthfully and without any attempt to misrepresent or conceal . . .

The Judge then referred to the attempts of murder made upon those individuals in high governmental offices of responsible authority, adding:

> I want to take the opportunity on this solemn occasion to say to the associates in crime of this defendant that they cannot by such acts terrorize American executives and prevent them from performing their plain duties, and they cannot prevent American courts from declaring the law exactly as they find it. Judges and executives may be placed out of the way by the hand of the assassin, but there will be others immediately to take their places just as ready and determined to perform their duties as their predecessors were, and backed by a public opinion that will ferret out and disclose the authors of every such crime.

Judge Wood pronounced the death penalty, but because of his conviction that Harry Orchard had testified to the truth in the interests of the state, he recommended to the Parole Board that his death sentence be commuted to life imprisonment. The Parole Board subsequently accepted the recommendation and commuted the sentence to life in prison.

The day of sentencing, Julian Steunenberg, the son of the victim, asked to see his father's assassin. Warden Whitney was somewhat surprised by the request and was a little apprehensive about the encounter. Orchard was informed of the request and agreed to a meeting. The two men cautiously shook hands as they measured each other. Then, instead of castigating the killer, young Steunenberg told Orchard to "Turn to God for forgiveness and the salvation of your soul." He then gave the convicted murderer some tracts and other publications of the Seventh-day Adventist Church. Strange as this meeting seemed, it was to have a decided effect on the prisoner. He was to carry the results of it with him to his grave.

CHAPTER 28

They may sentence Harry to be hanged a hundred times, but you cannot make me believe they will ever hang him.
. . . Ida Toney Orchard

The second trial of Steve Adams was the prosecution's last chance to force Adams to testify against the Inner Circle. Everything was based on their obtaining a verdict of guilty against him. His trial was slated to begin in October, 1907, in Rathdrum, Idaho, and one very important witness for the state was to be Harvey K. Brown of Baker County, Oregon.

Brown had been instrumental in securing the arrest of both Orchard and Adams and had testified in Adams' first trial. Since the trial, he had tried to convince C. W. Lillard, Adams' uncle, that he had made a mistake in persuading his nephew to repudiate his confession, but the attempt had been unsuccessful. For some time, Brown had been receiving anonymous letters threatening his life if he failed to leave the country. The former sheriff was also aware that his movements had been under surveillance for several weeks. His wife was breaking under the pressure. The situation was so straining that Brown finally decided to move to California.

On the night of September 30, Brown returned from inspecting his mining claims on Stice's gulch. A block away from his house, he noticed a man rapidly approaching him. Brown spoke, but the man did not respond as he continued hurriedly past Brown. Only a few days previous to this incident, Brown had remarked to a friend, "I'm a doomed man. The people in Idaho will get me yet."

As Brown opened the front gate of his picket fence, a hidden bomb exploded, throwing his body into space. The device had been placed just inside the gate on the south side of the walk. From the bomb, wire had been run across the walk, up to the picket fence and toward the road. From this position the explosive device had been detonated.

Brown survived just long enough to say, "It's over the Steunenberg case, I'm sure of it. I helped track down Steve

Adams, and ever since his friends have been on my trail, I believe." Then he died in agonizing pain.

Captain W. S. Swain assisted in the murder investigation, but he had little success, and the case remained unsolved. Brown was not the only law officer connected with the Orchard case to feel the threat of assassination. Three attempts were made on the life of Angus Sutherland, former sheriff of Shoshone county and the man who had arrested Vincent St. John in Burke, Idaho. Sutherland was duped by a fake telegram into going to Pullman, Washington, but was forewarned by a drunk on the train who revealed that three men were lying in wait for him at the train depot.

In October, Steve Adams' last trial in Idaho began. It was almost an instant replay of the first. Bulkley Wells testified that he first met Steve Adams on May 27, 1906, when he arrived in Idaho to transport the confessed murderer to Colorado. An attempt was made at that time to locate the body of W. H. Barney, the non-union shift-boss at the Smuggler-Union mine who had mysteriously disappeared and was believed to have been killed by Federation gunmen. During the trip, Adams told Wells of a number of crimes in which he had participated.

Before quitting the unsuccessful search, Adams insisted that he had located the exact spot. He claimed that the body had been buried in one place, and the clothes in another. Wells testified that he later made a rough drawing of the area and sent it to Hawley; he asked him to have Adams take the diagram and mark on it the spot the body could probably be found. Adams had made a sketch, and from that sketch they had finally located the body of Barney.

During the murderer's first trial, the body of Barney had not been found, and attorney Richardson made a point of that fact, trying to establish the validity of Adams' confession. Also, the testimony of Archie Phillips, the last man to see the murdered man, Tyler, alive, was read into the court record. Phillips knew Adams by the name of "Reddy," and he had been with Tyler when the accused and Jack Simpkins had fired a number of shots over their heads to scare them. The act was partially successful. Phillips and his wife were frightened enough to leave the country — never to return.

Darrow stated to the jury that Adams had been in North Idaho because Jack Simpkins had told him of the advantages of the Marble Creek district; and that he had gone there to see if he wanted to secure a timber claim for himself. It was shown that Adams used the name of Steve Dickson, the name of his wife's first husband, and that while he was liberally supplied with funds, the source of the money was vague.

The state, however, failed to break Adams' alibi that he was attending a birthday party at the time the murder was supposed to have taken place. As a result, the jury failed to reach a verdict and was discharged. They stood eight for acquittal and four for conviction, and so the State of Idaho dismissed the murderer. However, at the same time in Telluride, Colorado, a Grand Jury returned an indictment against Adams for the m rder of Arthur Collins in 1902. He was to be returned to that state to stand trial.

The Pettibone trial began on November 27, 1907, and was less spectacular than the Haywood ordeal. The defendant had been ill, and for a time it appeared that the trial would have to be postponed, but at Pettibone's request it convened. Little new evidence was introduced. Orchard had his day in court and performed adequately, but the public was too familiar with his story of terror. Harry Orchard was "old hat" to them. The state again was not able to corroborate his testimony, at least to the satisfaction of the jury and within the legal definitions set forth by Idaho law. George Pettibone was found not guilty, and subsequently the case against Charles Moyer was dismissed for lack of evidence. The state realized the futility of trying to get a guilty verdict.

On the date of March 10, 1908, the State of Idaho disposed of its last court case, that of Harry Orchard himself. The mass slayer was taken to Caldwell to be arraigned in the district court of Canyon County on the original charge of the murder of Frank Steunenberg. Previously, an accused person had not been allowed to plead guilty to a charge of first degree murder, but now it was permissible if the court allowed. Had Orchard stood on the plea recorded, not guilty, a trial would have to take place. Frank Wyman, the defendant's court appointed attorney, had no idea what his client was going to do at this point.

When he had been first arraigned on the murder charge after

his arrest, Orchard had stood mute before the court. The presiding judge had ordered that a plea of not guilty be entered into the records. Now it was the moment of truth for the confessed murderer, and he was seriously weighing all the alternatives. His partner in crime, Steve Adams, had made a confession, then repudiated it, and had "beat" two court trials. Why not the same for him? Anything could happen where twelve individual men were concerned, anything. Adams had already proved that point. Orchard had been advised he should plead guilty to murder two, so his mind must have been churning with the conflicting thoughts.

There were very few persons in the courtroom that day, an abrupt departure from the past. The court called the case of Harry Orchard. Attorney Wyman stated that his client wished to change his plea as it had been originally entered in the case.

Judge Wood asked the defendant, "You have heard the application of your attorney. Do you desire to withdraw the plea heretofore entered?"

"I do, your honor," was Orchard's firm reply.

"Your attorney has fully explained the law to you, so that you think you understand the consequences of the withdrawal of that plea?" asked the Judge.

"Yes, sir."

"What is your plea, Mr. Orchard, to the indictment upon which you have been arraigned?"

Silence. "My plea will be guilty, your honor." His answer was given with perfect calmness.

"The only plea that this court can accept is murder in the first degree as charged in the indictment," spoke the voice of the court.

"I will fix Wednesday morning, March 18, at ten o'clock, as the hour for pronouncing judgment in this case," said the Judge.

"I am guilty. I desire to take the punishment, and I am ready to suffer what the law demands for my crime," said a penitant Orchard.

On the date and at the time prescribed by the court, Harry Orchard stood in the center of the Canyon County courtroom. He stood motionless, his hands clasped behind his back, his eyes fixed on the face of Judge Wood. His face seemed paler than usual, and once his lips twitched — the only signs of emotion.

Surprisingly, the judge launched into a lengthy review of the entire case. Wood declared he had made a detailed investigation and had concluded that Orchard had never been promised any concessions or immunity. The defendant had testified, he stated, "fully and fairly to the whole truth, withholding nothing."

As the judge made his address, the killer's face took on a puzzled look, then it changed to one of surprise. When he was finished, Wood informed Orchard that he might make any statement that he wished before sentence was pronounced.

The defendant became confused, and any carefully prepared statements were forgotten. Instead, he said, "I want to thank you for all those things you have gone through so thoroughly." His voice softened. "I was promised nothing for my testimony." The killer paused, as if selecting his words. "I want to thank you and the kind, loving providence who has given me strength to go through with this." He took a deep breath and finished. "My only motive was to make all the amends I could for my wrongdoing. That is all I have to say."

The emotional testimony touched the people in the courtroom, and women were heard sobbing. However, Judge Wood firmly concluded his connection with the sensational murder case when he stated, "The judgment of the court, Mr. Orchard, is that you are guilty of the crime of murder in the first degree, and that you be punished by death inflicted by hanging by the neck until dead."

The sentence of death was to be carried out on Friday, May 15th. In pronouncing sentence, however, Judge Wood recommended that the Pardon Board, the final authority in the matter, commute the sentence to life imprisonment.

The sentenced murderer was returned to the penitentiary and placed under the death watch. All in all it had been a hell of a birthday for Harry Orchard — he was forty-two years of age.

Later that night the condemned man gave his permission for an interview. In a somewhat reflective mood, the mass slayer talked of the turbulent past. "When we blew up the Bunker Hill and Sullivan Mill, I didn't feel any particular responsibility for it. It seemed then as if we had a right to do that." He went on to talk of his wanderings through Utah, Nevada, Arizona and his arrival at the great gold-producing district of Colorado. "Finally, I reached Cripple Creek and there I met men whom I sympathized with up north. The fight was soon on there, and I readily took a side. I thought then that I believed we had rights that should be protected, by bloodshed if necessary. That's how it was."

Orchard raised his head slightly and sucked in a deep breath of air before he continued: "It wasn't what money I got for commiting the crimes entirely. I felt I was working for a cause." This was a more idealistic rationale for his participation in those bloody events than he previously had indicated.

In Cripple Creek, Ida Toney Orchard received the news of the death penalty imposed on her husband and snapped: "They may sentence Harry to be hanged a hundred times, but you cannot make me believe they will ever hang him." As an after thought she added, "He knows too much for the authorities of Idaho to put him in the grave."

Sherman Bell in Denver stated emphatically that, "Orchard will never hang." With the knowing wink of someone who knew a secret, he elaborated, "I know Orchard better than anybody, at least in some things. I am sorry that he will not be hanged, however, on general principles." As the old Rough Rider continued, he spoke with anger. "Why don't the state of Colorado try somebody as principals or accessories, either or both for the fifty or sixty Colorado murders? Why has it not been done?" The former adjutant general had raised questions that were on many minds.

In the meantime, Steve Adams was awaiting trial for the alleged killing of Collins, precisely the type of crime that Bell was talking about. Several days after Orchard was sentenced, Bulkley Wells returned to his home at the Smuggler-Union mine at Pandora and resumed his position as general-manager of the wealthy mine. Years earlier Wells had pledged he would bring the slayer of his predecessor to justice. Now, he had seen Adams indicted and brought back to stand trial for that murder.

On the night of March 28, after visiting friends in Telluride, Wells returned to the large frame building used for offices and sleeping quarters. It was midnight and there were no other occupants in the building at that time. His family was staying in

Colorado Springs. At twelve-thirty he retired to an open porch over the dining room and quickly dropped off to sleep.

About two o'clock, with a terrifying force, a bundle of six sticks of dynamite exploded under his bed. Wells was thrown violently fifteen feet across the room and buried among debris; he was unconscious, but alive. Although he bled profusely, his only serious injury was impaired hearing. Since the force of the exploding dynamite had been downward, Wells had miraculously survived. Gaining consciousness, he staggered to the telephone and, barely able to speak above a whisper, he managed to summon help.

Ironically, the murder attempt had taken place in the same house in which Arthur Collins had been gunned to death by Steve Adams. Harvey K. Brown had been assassinated just before the second trial of Adams, and now Bulkley Wells had been the target just prior to the killer's trial in Grand Junction.

On April 3, Frank T. Wyman, Orchard's attorney, applied for a commutation of the sentence given to Orchard. His client absoluely refused to consider the legal move by reiterating his desire to die. Under Judge Wood's direction, though, Wyman filed with the state board an application for pardon. Wood did not want Orchard to die. He was far too valuable a witness to lose. As the judge said at that time, "Should Harry Orchard be put to death there would be no reason why Jack Simpkins might not walk the streets of Caldwell or any other Idaho town with immunity."

On July 1, 1908, the State Board of Pardons, by a vote of two to one, commuted Orchard's death sentence to life imprisonment. Governor Frank Gooding and Attorney General J. J. Guheen voted in favor of reducing the penalty. The Secretary of State voted that the original sentence of the court be carried out. It appeared as if Harry Orchard had cheated the gallows, after all.

On hearing the board report, Warden Whitney so informed the condemned man and reported the killer's displeasure with the news. Orchard had been reconciled to his fate with the hangman, and he opposed all efforts of his attorney, who considered his action his legal and moral duty as a lawyer. The preparations for Orchard's execution had been made, including the preparation of the ropes that would be used to hang him.

The scaffold which he cheated was located just inside the prison walls on the east end of the institution.

Orchard settled down to a lifetime in prison. Meanwhile, his former partner in crime, Steve Adams, was being tried in Colorado. On July 15th, the twelve man jury agreed that Adams was innocent of the crime and acquitted him. The confession which Adams had made while he was confined in the Idaho prison was ruled by the court as not admissible as evidence. The defense had produced eight witnesses who testified that Adams was playing cards that November night when Arthur Collins was shot to death. Adams supposedly was still in the card game when a man brought the news of the killing. After the jury reached its verdict, Adams left the courtroom with his old friend and fellow warrior of the labor wars, Bill Easterly. Later that day his friends escorted him to the westbound Rio Grande passenger train bound for Baker, Oregon, and his home.

After his acquittal, no move was made to arrest Adams on any of the other murder charges connected with the state's labor-troubles. Authorities were now convinced of the impossibility of their securing a conviction. They noted that Idaho also had that same problem. Reluctantly, the State of Colorado rested its case.

In July, 1908, the Western Federation of Miners held its annual convention in Denver. Charles Moyer made an address in which he urged the use of peaceful methods to reach working agreements with employers. The president stressed that he deplored the calling of unnecessary strikes and recommended changes in the Federation's constitution to require local unions to consult with the executive board before they ordered a strike.

Moyer was a changed man. He had taken Bulkley Wells' advice and had decided to become more conservative in his approach to labor affairs. And as a result, the convention was a power struggle between the forces of Moyer and those of Haywood. When Haywood spoke to the gathering, Moyer walked out. The extreme radical element was through in the Western Federation of Miners — a change was in the wind. The organization would never be the same; Moyer would personally see to that.

With the conclusion of the Haywood trial, the "death rattle" of the Western Federation of Miners was sounded. Public disclosure of the violent methods used by the organization produced results like gonococcus bacteria being exposed to penicillin. The fact that labor leaders had received justice in the Idaho courtroom was a blow to radical socialists who had contended, in preaching the class-warfare doctrine, that a fair trial for them was an impossibility in a capitalist society.

The Federation slowly began evolving into a more conventional, labor-union structure. In 1916, the name was dropped, and its new identity now was the International Union of Mine, Mill and Smelter Workers. As for the violent past of the organization, that era was over and done with. Of the rampant and turbulent men who were associated with Harry Orchard in his madness, only he would be legally accountable for the crimes committed. Haywood, Moyer, Pettibone, Adams, Easterly, Parker, Simpkins, St. John, Minster, and Davis all escaped courtroom justice. What happened to these men and others who were so closely associated with Harry Orchard and his crimes? How did they fare while the murderer spent the rest of his life in prison?

CHAPTER 29

By The Grace Of God

One of the most fascinating aspects of the Harry Orchard case was the personalities of the people whose lives he so significantly touched. The violent labor problems that plagued the mining districts of the Western United States produced strong-willed men on both sides of the disputes. They had to be strong in order to survive. Blood ran freely, and weaklings were quickly eliminated.

In the fantastically wealthy mountain region of the Coeur d'Alenes, Harry Orchard's one time partners continued to mine for the elusive rich ore with fanatical zeal. Time, effort, and money were lavished on an enterprise that other mining men, including experts, proclaimed as the "useless pursuit of fools." But to the group of amateur miners whose persistent drive enabled them to discover high grade ore in the depths of the workings they called the Hercules, the pursuit was part of an American dream.

Discovering high grade ore was only the beginning of the ceaseless struggle. Next came the struggle to develop their property without yielding to outside control. Every step of the way they had to fight the interests of the big mining companies who quickly moved to grasp the property. They faced the almost insurmountable task of financing the project, and hard rock mining takes money. Nonetheless, the small group continued, overcoming many obstacles. They eventually succeeded and became millionaires.

The Day brothers — Harry, Jerome, and Eugene — on the strength of their success with the Hercules, developed other mining properties and earned an enviable reputation for their astuteness. The family became a political power in state politics, but only Jerome actively entered the field by serving as a state senator. Eugene died in 1922; Jerome in 1941; and Harry in 1942. The Day family is still represented in the mining industry in the great Coeur d'Alene district by Henry L. Day, son of Harry, now president of Day Mines, Inc. The Hercules mine itself, a cherished legend in mining lore, was forced to close in 1925, its once rich ore veins diminished.

Another of the successful operators was August Paulsen, the Danish immigrant, who moved his family to Spokane, after the wealthy Hercules mine began paying dividends. He, too, proved to be an astute business man and built the city's first skyscraper. When Orchard's confession became known, particularly the details of his plans concerning the proposed kidnapping of the Paulsen children, a grim pall had descended over the family. Paulsen no doubt remembered better times, before his friend had become a professional assassin and before the bloodletting had begun. He died in March, 1927.

After the trials the powers in the Western Federation of Miners were free from legal action with the possible exception of Jack Simpkins, who mysteriously disappeared and was never located. Law officers all over the western part of the United States made determined attempts to find him. He was reported to have been in a dozen places, but attempts to confirm his presence always failed. Simpkins' whereabouts would always remain a mystery.

George Pettibone had little time left on earth after the Boise jury brought in the verdict of not guilty. A cancerous growth around the kidneys had bothered him before his trial, and he died in August 1908. John O'Neill, editor of *Miner's Magazine*, believed that Pettibone's health was ruined by his long confinement in the cell below the surface of the earth in Boise. Pettibone had been made an honorary member of the Federation and always was popular with the rank and file of the union. Contrary to Orchard's prediction that Pettibone would make a death-bed confession, it never came. He died, as he had lived, a strong supporter of the Western Federation of Miners.

Charles Moyer died of asthma during his retirement in Pomona, California, in June, 1937. At one time the announcement of his death would have brought a sigh of relief from the leaders of the mining industry and cries of grief from the miners. At the time of his death, one of his chief antagonists in the labor wars, Sherman Bell, was an invalid at Fitzsimmons General Hospital in Denver.

Moyer had become a conservative in his labor relations. By the 1920s even James Hawley praised his leadership, although he was condemned by the more radical elements in the labor movement. Nevertheless, no man was as responsible for the Federation's joining the ranks of the American Federation of Labor or adopting a new name as was Charles Moyer.

Big Bill Haywood's militant attitude was too radical for the strife-ridden Federation, so they expelled him from Denver. His connection with the union was officially severed when the following notice appeared in the *Miner's Magazine:*

> Notice. To whom it may concern: This is to inform you that the Executive Board of the Western Federation of Miners has decided to terminate the services of William D. Haywood as a representative of the Western Federation of Miners in the field, the same to take effect on the eighth day of April, 1908.
>
> C. E. Mahoney, vice-president, W.F.M.

Not to be outdone, the Socialist Party removed Haywood from his position as executive board member and deprived him of further party membership. By 1913, even the violent Haywood believed the old labor wars of the Cripple Creek and the Coeur d'Alene days were a thing of the past and that the extreme violence waged at that time had accomplsihed little.

This dynamic personality was just reaching the zenith of his career when he became the secretary-treasurer of the I.W.W., succeeding his old friend and comrade, Vincent St. John. In a few short years Haywood was to become the best-known leader of the "Wobblies," and as such his reputation became international in scope. He was a forceful speaker, a brilliant organizer, and a powerful persuader. In September, 1917, the big man was arrestd and convicted under the Federal Espionage Act. In 1920, while he was out of Leavenworth Penitentiary on bail, he fled the country, convinced that returning to prison would shorten his life. He was suffering from diabetes, failing eyesight, and above all he lacked the spirit to continue the fight.

Convinced by friends in the Communist Party that Russia was the land of opportunity and would instantly recognize his genius, the labor-leader sought sanctuary there. Haywood met Lenin and was sent to supervise the Nadjedinski Iron Works and the coal mining district at Kemerov. Viewing first hand the actual workings of Communism, he became disillusioned and bitter. Haywood, now a very sick man, returned to Moscow and was relegated to becoming a showpiece for Communism. Before his death he married his Russian mistress. He died in May 1928, after suffering a stroke. Cremated, his ashes were divided and placed in the Kremlin walls and in Chicago's Waldheim Cemetery, near the graves of the Haymarket martyrs, an incongruous ending for the hard-rock miner from Silver City.

After the Colorado murder trial, Steve Adams dropped out of sight somewhere in Oregon. When he died in December, 1934, a local Spokane paper reported the death of the professional assassin who with Harry Orchard had pulled the chair rung that set off the dynamite that destroyed the Independence depot:

> Steve Adams, 62, was found dead in his two-room home at Latah Saturday by sheriff's officers investigating a phone call. Coroner Barhart accompanied the officers and blamed the death on natural causes.

Vincent St. John went to Goldfield, Nevada, when he left Colorado; he became a leader in the I.W.W. during the miners' strike of 1907. He was then elected secretary-treasurer of the I.W.W., a position he held until 1914. A popular leader of the "Wobblies," he was referred to as the "Saint." St. John was convinced that political action was useless and that direct economic reprisal was the only way to affect change. St. John left the I.W.W. to become a prospector in New Mexico, where a chronic bronchial condition contributed to his death in 1929. This condition had developed after he lead a rescue party in a mine disaster in Telluride.

William Edgar Borah, along with Clarence Darrow, became the best known of the participants in the Haywood trial. Strangely, Borah's fame eclipsed the excellent work of the chief prosecutor, James Hawley, and he forged ahead to become a leader in the United States Senate for almost thirty-three years; he supported many progressive reforms such as the income tax, direct election of senators, national prohibition, and labor legislation. Borah is probably best remembered for his opposition to America's becoming a part of the League of

Nations and the World Court. One of his many supporters and great admirers was his friend and legal contemporary, James Hawley. The rare friendship between the two men was a strong bond that lasted until Hawley's death. Borah, the "Lion of Idaho," died in his Washington D.C. home after a stroke in 1940.

The Haywood trial gave Clarence Seward Darrow, as in the case of Borah, national exposure and insured him a place in the legal profession that became almost legendary. The "labor lawyer" surged on to become best known as the defender and legal champion of the unfortunate and the oppressed. Darrow brilliantly defended Loeb and Leopold in their 1924 trial for the murder of young Bobby Frank. In the 1925 famous Scopes trial, he defended the right to teach evolution in the Tennessee schools. In later years the attorney toured the country, debating social issues with prominent opponents. When he died in 1938, he had the reputation of being the nation's greatest criminal lawyer.

The voters of the State of Idaho, in gratitude for his years of service, elected James Hawley governor of that state in 1910. An indication of his popularity was the fact that he was the only Democrat elected on the state ticket. One of Hawley's dreams was to be elected to the United States Senate, but it was never fulfilled. He was the first lawyer to be elected governor, and because of his popularity he was known as "Idaho's Grand Old Man," until the time of his death in 1929.

Edmund F. Richardson, the brilliant defense attorney, returned to Denver at the conclusion of the Haywood trial and after his break with Darrow. The trial seems to have been his last connection in a legal way with the Western Federation of Miners. His personal opinion of the defendant, Haywood, and his continuous disagreements with Darrow seemed to weigh heavily on him. In the end, he concluded that Bill Haywood had been more involved in Orchard's activities than he had previously suspected.

On May 7, 1911, while he was driving on the motor-road between Denver and Boulder, Richardson swerved his automobile to avoid a collision with another car on a hazardous curve and lost control of his vehicle. The car plunged over the edge of a steep enbankment, turned over one and a half times, and finally rested at the bottom of a ravine on its right side. Mrs. Richardson, their three children, and another passenger were thrown clear of the wreckage and escaped injury, but two men were pinned under the huge Thomas touring car. Richardson and Horace Granfield, a house guest, were fatally injured. Ironically, Granfield was a millionaire mine owner with financial holdings in the Cripple Creek district.

John Nugent, the defense attorney from Silver City, was appointed by the Idaho governor to fill a vacancy in the United States Senate in 1918. The other Idaho senator, William E. Borah, had been his opponent in the Haywood trial. Nugent resigned his Senate seat in 1921, and again a governor filled the vacated position by appointment. This time the man to occupy the Senate office was none other than the former governor, Frank Gooding, who died in that office in 1928.

James McParland, acknowledged to be one of America's finest criminologists, died in May, 1919. The detective occupied a special place in two of the most infamous murder conspiracy trials ever held in the United States. The Haywood trial was the last case of national scope for McParland. The professionalism he demanded from his agents is best recorded in the files of the Pinkerton Agency.

McParland died forty-three years after the Molly McGuire trial had been held in the Pottsville, Pennsylvania, courthouse. He was best known for his part in bringing those murderers to justice, a fact that alone insured his place in the annals of detective lore. McParland always believed that Orchard merited release from prison because of the testimony he had given in the Haywood and Pettibone trials. but that opinion was not shared by his employer, William Pinkerton, who refused to endorse any such recommendation.

Probably the most tragic end of any of the figures associated with Harry Orchard was that of the dapper militiaman, Bulkley Wells. In the years following his clashes with the Western Federation of Miners, Wells became wealthy and was recognized as one of the foremost mining and hydroelectric engineers in the country. When he divorced his wife, Wells suddenly became the most eligible and sought after bachelor in Denver society. On one of his numerous business visits to San

Francisco, he met the wife of a friend and associate, fell in love, married her, and disappeared from Denver society forever.

By the early twenties, Bulkley Wells had left Colorado for the West Coast, embarking on a gigantic mining development program. Wealthy himself and backed by the millions of Harry Payne Whitney, he organized a competent engineering staff with headquarters in San Francisco and launched a series of mining ventures. Wells gained possession of the United Comstock property near Virginia City, Nevada, and constructed a cyanide mill there. At Gold Canyon, near Silver, Nevada, he built a dredge and reopened old mining property at Grass Valley, California. Altogether, more than a dozen properties were brought into the Whitney-Wells organization.

Then disaster struck. The United Comstock collapsed because of a drop in the silver market. Wells bought mining property in northern Mexico, but realized nothing. The magic touch seemed to have disappeared — available silver was worthless, and he could not amass enough gold to recoup his fortunes. One by one his business interests failed, followed by numerous disputes and claims over mining property which was tied up in courts. In fighting for his financial life, he attempted to recoup his losses by playing the stock market. He failed there, too.

The combination of events was too much for the man who had dared to ride the streets of Telluride mounted on a white horse, "A target that none might miss." He was a tough individual who could stand up to big Bill Haywood. He could lose a fortune and never change the expression on his face, and he could tell the leaders of Denver's high society to go to hell. But now he was dead broke, and his pride was withered. One day in May, 1931, depressed and broken, he asked a friend who shared an office with him for the loan of twenty dollars. The friend left for the bank to withdraw the money. Ashamed, Bulkley Wells placed a pistol to his head and slowly squeezed the trigger.

In the meantime, what was happening to Harry Orchard? On January 9, 1909, he was escorted under guard to the prison's big plunge bath. In the presence of witnesses, he was baptized into the faith of the Seventh Day Adventists. He had studied the religious material the Steunenberg family had given him and had made a decision. No doubt the interest the Steunenbergs had shown in him was responsible for the murderer's conversion. Perhaps he felt this to be a small measure of atonement for his taking of human life.

The wife of the slain governor continued to send the assassin religious materials to strengthen his belief in God. She genuinely felt that this man's soul could be saved. When Orchard learned that Mrs. Steunenberg wanted a confrontation with him, he was terrified by the idea. The murderer wrote and begged her not to come. Nonetheless, they finally met behind the solid, grim prison walls. Harry Orchard, who could kill without a trace of emotion for his innocent victim was overcome with grief and shame. The widow, too, suffered because of her concern. The public could not understand such actions, and some family members resented it. But strange as it may seem, an unusual relationship was formed, one that would survive for decades.

In the spring of 1909, John W. Snook became warden of the prison. Snook had been a deputy United States marshal, in the Territory of Alaska during the Klondike gold rush, and was a big, aggressive man with a good deal of common sense. He found the most celebrated prisoner in the institution in solitary confinement. Snook lost no time in paying convict number 1406 a visit, and they held a long, detailed conversation. He discovered that the last thing Orchard wanted was to be released from the safety of the penitentiary. To walk out the front gate would be a signature on a death warrant for him. Being safe was to be in legal custody, surrounded by armed guards.

At the conclusion of the conversation, Warden Snook informed Orchard, "You're not going to be kept in a cell here. You're going to be treated like every other prisoner." The prisoner was then taken to the general cell house and began eating his meals in the main prison dining hall. Eventually, Snook allowed Orchard to work outside the prison walls and behind the horse barn. His job was to care for the large turkey herd which supplemented the institution's food program. Prisoner number 1406 never betrayed that trust.

Most of the money Orchard received from the McClure Company in payment for his autobiography, published in 1907, was sent to his wife, Florence. At times, to meet his wife's financial needs, the prisoner borrowed money from his wardens, debts which he always repaid. Realizing that Orchard was supporting his family, Snook allowed him to make small articles and objects which he could sell. When Warden Snook left his position in 1917, Orchard made him a fine grade hair brush as a gift of appreciation.

For many years the murderer worked in the prison's shoe shop, which he had been given permission to start. Borrowing twenty dollars from the warden, Orchard bought sole leather, skins, and other supplies. The prisoner first created bench-made shoes for state officials, and when the demand increased, special orders were filled for the public. During World War I, he made a harness set for the former governor, Frank Gooding. The leather used was the best, and the harness was completely hand stitched. He received three hundred dollars for his labor. Expanding his shop, the enterprising prisoner purchased stitching machines on credit. As his business grew, sales representatives regularly called upon him at the prison with orders for merchandise, just as in any business.

The lucrative shoe shop business ceased with a change of wardens. Since there was little work for the majority of the inmates to do, the prison board decided that the shoe shop should be managed by the institution and that it should pay for itself much in the same manner as Orchard had operated it. No state appropriations were made to support the operation, however, and the only way the prison board could operate it was to assume that it would stay in the red. The operation was transferred into a large building, intending to expand the business. They confiscated all of Orchard's personal machinery and agreed to pay him, for a period of time, a royalty on all the work done in the shop. Later the state legislature authorized the payment of several thousand dollars to the mass slayer, and the board then assumed complete responsibility for the enterprise. The convict capitalist was out of business.

In 1920, prisoner 1406, then fifty-four years of age, began thinking seriously of life outside the prison walls. The Federation was no longer organized; he had reformed and turned to religion; and he had been a perfect inmate. To pave the way for a successful parole hearing, he began to line up support from among the more important persons he knew. McParland was dead, but he realized that a strong endorsement from the Pinkerton Agency would have even greater impact. Orchard sent a request to William Pinkerton in Chicago, but the gentleman refused to have any part of it, still considering Orchard nothing more than a professional assassin.

In 1922, however, Harry Orchard felt the time was right for him to apply for a parole. He had an unusually impressive list of persons who were ready and willing to recommend his release. The bid for freedom came on November 13, 1922, when his application came before the Pardon Board. Quite possibly no other convict in the United States had ever had such a prestigious group of individuals pleading his cause. Among them were United States Senator Frank Gooding, former governor James Hawley, former Justice of the Idaho Supreme Court, I. N. Sullivan, and F. A. Jeter, Idaho Secretary of State elect. Each of these men stated he was not there to represent Harry Orchard, but he considered it his duty as a private citizen to endorse the parole movement. The group contended that the prisoner had rendered a great service to the state, and, for that matter, to all the mining regions of the West. Orchard's confession had sounded the "death knell" to the Western Federation of Miners, finishing it as a criminal organization.

Senator Gooding related that he had discussed the Orchard case with William J. Burns, head of the United States Secret Service, who had agreed that the prisoner should be given his freedom. His reasons were that the rewarding of criminals who turned state's evidence would do more to secure convictions than anything else. And James Hawley, the old war horse, emphasized that English common law, which is applied in this country, provided a reward in the form of leniency to a prisoner who gave information which aided the prosecution.

Support came from all over the United States. Charles Siringo, the cowboy detective wrote, "It is my sincere wish before I die to see Harry Orchard freed from prison." The Reverend Edwin S. Hinks called Orchard a "benefactor to humanity." Even Bulkley Wells said he believed the prisoner had sincerely repented of his criminal acts and had served a term sufficient to represent atonement. Publisher S. S. McClure claimed, "I think it would be wholly right to release him, and I go further and say that I think it would be wrong to keep him."

Without a doubt the most unusual request came from the victim's family itself, in the following signed statement:

"We, the family of the late governor Steunenberg of Idaho, do not object to the release of Albert E. Horsley, better known as Harry Orchard, from the State Penitentiary.

Eva Steunenberg
Julian Steunenberg
Frances Steunenberg
Frank W. Steunenberg"

Few, if any persons appeared before the board who objected to the release of the prisoner. Harry Orchard might have been released had it not been for one particular incident.

When the news of Harry Orchard's application was made known, Charles B. Steunenberg, brother of the victim, decried the action in an open letter to the *Boise Statesman*. Mr. Steunenberg wrote that he had not been notified of the hearing even though several times he had requested that he be informed. By composing a classical retort he challenged the speeches made before the board:

> The argument is also set forth that Harry Orchard is a changed man. This is undoubtedly true, and the way to assure society that he will continue to be a changed man is to keep him right where he is so that he has no opportunity to be otherwise.

The Steunenberg letter accomplished its purpose — it stirred public protest. Citizen reaction was that the public would not tolerate such a move, and consequently the pardon was not granted. The chance of a lifetime had been lost due to a single dissenting letter. All over the country, certain individuals relaxed at the end of the hearing, especially those who had been marked for death by the assassin. Harry Orchard accepted the rejection philosophically; to him, it was the will of God.

In the middle of the 1920s, the mass murderer asked the prison warden for permission to start a poultry farm on the property at the rear of the prison. His request was granted, and Orchard began this venture by investing in several hundred baby chickens. Later, he built two poultry houses — one to accommodate two thousand hens, and the other to quarter more than a hundred. The venture was a winner from the start, and the former assassin poured his energy, time, and devotion into the project. So that he could watch over the growing number of chickens, special arrangements were made for Orchard to erect a small cabin outside the prison walls. It consisted of two small rooms, and it was filled with furniture he had made himself. This abode was probably the most comfortable surroundings any "lifer" enjoyed in any penitentiary in the United States. He had everything except freedom.

One of Orchard's frequent visitors, whenever he happened to be in the area, was the incomparable Bulkley Wells. On one occasion when the mining engineer was visiting, Orchard mentioned he desperately needed a typewriter for business purposes. Wells left the penitentiary, and within a few hours returned carrying a new machine which he presented to the convict. A few years later, when Wells was again visiting the old dynamiter, they discussed Orchard's needs. He remarked that he could use a lathe in his work. Despite the fact that he was already having financial difficulties, Wells purchased the lathe. This was to be the last visit of the former militia officer. Several years later, he was a suicide.

As decades passed, chances of Orchard's being released grew dimmer. Senator Borah's remark, made during the Haywood trial, that the State of Idaho had no intention of ever allowing the confessed murderer to be free, began to have the ring of truth. Ironically, the killer's closest confidants were the same men who had captured, prosecuted, and used him. Orchard showed as much emotion over the deaths of McParland, Gooding, Wells, and especially Hawley, as he was ever capable of exhibiting.

During the years, his religious devotion increased. He learned how to express himself in prayer, and pray he did — morning, noon and night. One day while he was working in the commissary, another inmate was surprised to find Orchard praying during the lunch hour. The prisoner took one look at the kneeling mass murderer and decided the old man was delirious. Not so!

When the state took possession of Orchard's shoe business and moved it to a larger building, the prisoner asked if he might use the old building to conduct prayer meetings. Thus, Harry Orchard the slayer of many innocent men, became the leader of a Bible study group. He became fond of saying, "By the Grace of God." However, when Orchard began offering to his church tithing on his earnings, the warden argued that any money he earned should go to his wife and his child. Orchard was adamant, however, and the church received the tithing.

Most of the prison personnel who worked with the celebrated criminal over the years considered him absolutely trustworthy. Orchard was described as a model prisoner. There were personality clashes and differences of opinion, to be sure, but he proved to be a man of his word. Because of his status as a lifer and his length of service, he had privileges that other men did not share. During the thirties, one warden in particular considered inmate number 1406 as disagreeable and abused privileges as if he ran the institution. He believed that Harry Orchard had not basically changed over the years and was shocked at the unconscious display of sadistic delight when Harry Orchard killed the turkeys and chickens. But the same warden admitted that the prisoner was a diligent worker who attempted nothing illegal.

While in prison, Harry Orchard wrote several autobiographies. The first was written immediately after his confession and was an enlargement on that document. He finished another in the 1940s at the age of eighty-two, but he was unable to get it published. Disappointed, he packed four manuscripts, diaries, and letters and sent them to an old friend, LeRoy Edwin Froom, an official of the Seventh-Day Adventist church, asking that they be corroborated and published. The resulting work appeared in 1952 and was appropriately entitled: *Harry Orchard, The Man God Made Again*. The book is used by the church as a testimony to what religion accomplished for a murderous man who had a "fatal" fascination for explosives.

Warden Lewis Clapp had some reservations on the subject of murder, but Orchard reiterated time and again to him the injustice of America's honoring her armed forces for killing enemies during the world wars, when he had been sent to prison

for killing capitalists who had waged an undeclared war against the rank-and-file of labor.

Throughout his years behind bars, although Orchard was basically a loner, he felt a sincere obligation towards younger convicts, especially those who were first offenders. He seemed to be convinced, in his old age, that the Lord had spared his life so that he could be used as an example to other men who might be headed toward a life of crime. The newer inmates were impressed. Here was Harry Orchard, the professional assassin, a name known to almost all Idaho school boys. It was almost unbelievable that this old, white haired, round man with the ruddy cheeks, dressed in standard prison garb could have committed the gruesome murders about which they had heard so much. That was one thing Orchard never denied — that he had been a hired killer for the Western Federation of Miners. Some of the prisoners he did reach, and perhaps he affected their lives. Some were awed, while others just tolerated the religious old "bastard."

A persistent rumor that circulated inside the prison was that Orchard had refused a pardon, a chance to be free. This was not true. He was never offered a release. Although he was asked many times by journalists if he wanted a pardon, his answer was always "No." Prison had become his home; he would never leave it alive. In the end, he did not want to leave, for it had become his life.[83]

The murderer realized that his wife and daughter suffered enough because of his actions, and he wanted to cause them no further publicity and embarrassment. He refused to discuss them with the press. When he knew he would never be pardoned, he was faced with an inevitability that was extremely painful to accept. When his wife died, he had only his church to which he could turn, and he further immersed himself in its religious teachings.

In July, 1950, the murderer's small house burned to the ground. Lost were the small items that had meant so much to the old man — letters, pictures, and other personal remembrances. Even his good health began to fail. He was moved inside the walls of the prison and assigned a final job — tending the pumphouse on the graveyard shift. His one big passion was to indulge his appetite for ice cream. Then came the partial stroke that hospitalized him. Harry Orchard was a sick, elderly man who knew the end was near, and perhaps he welcomed it. By this time, he had even grown tired of answering questions about his criminal career. Replying to the questions of journalists, he would say in a tired voice, "Read my books — it's all there."

The last year of his life was a long one. Hospitalized, he was senile, hampered by a speech impediment, and unable to fully care for himself. He had no close friends except his church's officials. His body had wasted and shriveled. The guns of the industrial war had long been silenced. The murderer was the last weapon to go; he was a museum piece. He had, however, taken the precaution to insure his proper burial by signing legal papers several years earlier, giving all his possessions and the money from his books to the church. The mass murderer had completed his regeneration and was ready to meet his maker.

EPILOGUE

As the hands of the prison clock struck 8:09 on the morning of April 13, 1954, the old man's heart ceased to function. It had pumped life blood for eighty-eight years, but it was now stilled forever.

Accompanied by stern, business-like guards, the morticians were led to the infirmary. The physical remains of Harry Orchard, prisoner number 1406, were covered by a sheet. The body was examined briefly by the coroner, and with respect for the dead, the frail body was wrapped in a gold-rubberized blanket and removed with care and dispatch.

Funeral services for Harry Orchard were held in a Boise Seventh-Day Adventist church. The rites were conducted by Dr. Leroy E. Froom, who had witnessed the baptism of the dead man. On this day he testified to the regeneration of the murderous man whom "God had made again."

Harry Orchard was interred in the city's highly respectable Morris Hill Cemetery. Some distance from Orchard, across numerous headstones and grassy areas, lay buried prosecuting attorney, James Hawley. Albert Horsley alias Harry Orchard, was at last free from his earthly prison.

Notes

[1]His parents were Henry and Margaret Horsley. An 1871 personal census of the Brighton Township lists Henry Horsley, age 35, Church of England, English, servant. Margaret, age 29, Wesleyan Methodist, Irish. All of the Henry Horsley family were born in Ontario.

[2]Peter McKinney, a relative on his mother's side of the family, was to hang himself apparently because of family troubles.

[3]The child, Olive May Horsley, was born May 9, 1896, in the Brighton Township, Northumberland county.

[4]An investigation of the fire was held, but nothing was charged. Orchard claimed he received eight hundred dollars in settlement.

[5]During the Haywood Trial, defense attorney Richardson questioned Orchard about the wife he had left behind in Canada.

Q. "You never sent her a cent up to that time in all the years you deserted her. Isn't that true?"
A. "It is true."
Q. "And she had to work like a slave to live?"
A. "She worked in a cheese factory making cheese, so I understand."
Q. "And you know that is mighty hard work, isn't it?"
A. "Not for a man."
Q. "But for a woman?"
A. "It is too hard work for a woman," replied Orchard.

[6]The first union to organize in the Coeur d'Alene district was the Wardner Miners' Union, in November, 1887.

[7]George A. Pettibone, a native of Erie County, Pennsylvania, was an unusual man. Sandy haired, athletic, and robust, he could be sharp-tongued in conversation. Descending from revolutionary stock, he became a wanderer and canvasser selling household goods for the American Wringer Company. Gradually he worked his way to Montana, operating a saloon in the mining camp of St. Mary called "The Bucket of Blood." From there he drifted into the Coeur d'Alene mining area and became involved with union organizations. Pettibone assisted in forming the Central Executive Committee of the Miners' Union which became the real power behind union activities. Joseph Pinkham, United States marshal in Idaho in 1892, escorted the fierce George Pettibone, convicted prisoner, to the Federal prison at Detroit, saying this about his official charge: "I had an opportunity to study his character pretty closely, and he impressed me as a man to have no stamina or backbone; as soon as he was arrested, he became exceedingly docile."

[8]As Ed Boyce, one of the prisoners related the incident, "We were in conversation in front of the window, when Mr. Hawley stated that without a unification of the different unions we were practically helpless."

[9]The Tiger Hotel in later years housed the employees of the Hecla properties in Burke. A highway ran under one part of the building and a railroad under the other. A creek was also channeled under the structure. In steam locomotive days, it was necessary to close the hotel windows before the train passed through.

[10]On August 24, 1889, Harry L. Day and Fred Harper filed location notices on the Hercules and Fire Fly claims. The claims contained a small exposed streak of

quartz with traces of lead carbonate and iron oxide. On face value, the claims did not seem to offer a very wise investment prospect.

[11]Share owners either did manual labor in the mine or paid their pro rata share each month. Records for the Hercules Mining Co. for the months of May and June, 1898, show that Orchard paid $53.50 into the company.

[12]The facts concerning the selling of the mine interest to Cardoner and the registering of the deed on March 7, 1898, became important features of the Haywood trial. The defense tried to prove that Orchard lost his mine interest as a result of the labor riots of 1899, and because of Governor Stuenenberg's part in the happenings, Orchard murdered him. The truth was that Orchard had sold all his business holdings prior to the disturbances. Ironically, by selling the mining interest, he missed becoming an extremely wealthy man. The actual discovery of the Hercules ore body was made in the No. 2 adit, some 2,000 feet from the surface in June of 1901. Six hundred feet of cross cut had been excavated when August Paulsen set a powder charge and broke into a cave of silver and lead carbonate.

[13]The next time he would see McHale and Flynn would be at the Haywood trial when they both testified that Orchard was playing poker with them and not blowing up the Bunker Hill & Sullivan Mill as he claimed.

[14]Charles Siringo claimed that George Pettibone told him that the power of the Central Union of the Coeur d'Alene district was such that they could hire men to kill or to destroy property. The local unions were required to pay money to the Central Union when notified that a certain sum was to be spent for the "Good of the Cause." No records were kept showing the disbursement of this blood money.

[15]Prominent members, at one time or another, in the Coeur d'Alene Central Union were George Pettibone, Edward Boyce, L. J. Simpkins, and Marion Moore.

[16]A Mr. Patterson worked for the Robert S. Roe Transfer Company of Denver during the late 1890's. He claimed to have hauled a lot of rifles which were packed in piano boxes from George Pettibone's store in that city and were sent to the Coeur d'Alenes mining district.

[17]Harry Orchard later explained away the events of April 29th: When we blew up the Bunker Hill and Sullivan Mill, I didn't feel any particular responsibility for it. It seemed then as if we had the right to do it."

[18]Records of the Tiger-Poorman Mine for May, 1899, show that Harry Orchard worked three shifts and earned a total of $10.50.

[19]Orchard talked with William Haywood about the destruction of the Bunker Hill and Sullivan Mill and union activity in that labor dispute. In his opinion, Orchard told the labor leader, the whole affair had been a mistake since it spelled the finish of the Federation in the Coeur d'Alene mining district. Haywood's rationale of the entire event took a different track. To him the action on that April day was one of the main reasons for the increase of membership in the Western Federation of Miners. The union could well afford to lose the Coeur d'Alenes because of the large gains they made in Colorado, Utah, and other sections of the country. The blowing up of the Bunker Hill and Sullivan Mill was the method adopted to remind people of the power of the Federation.

[20]The application for the permit follows:

Application for leave to seek employment in the mines of Shoshone County.
To Dr. Hugh France, State Representative:
Sir: — I hereby make application for issuance to me of a permit allowing me to

seek employment in the mines of Shoshone County.
I am a by occupation.
I am a native of and am a citizen of the United States.
I worked at the mine in
My shift boss was
Heretofore I have been a member of Miner's Union.
I did not partake actively or otherwise in the riots which took place at Wardner on the 29th of April, 1899. Believing that the crimes committed at Wardner on that date were actively incited, encouraged, and perpetuated through and by means of the influence and direction of the Miner's Unions of the Coeur d'Alenes, I hereby express my unqualified disapproval of said acts, and I hereby renounce and forever abjure all allegiance to said Miner's Union, of which I was a former member, and I solemnly pledge myself to obey the law and not again seek membership in any society which will encourage or tolerate any violation of the law.
Dated this day of 1889.

[21]In his testimony Corcoran stated that he had arrived at the union hall at eight o'clock on the morning of April 29 and remained there until eleven, during which time he examined the books. He then returned home for lunch and stayed until four in the afternoon. A witness, however, stated that he definitely saw Corcoran sitting on top of a boxcar on the train trip back from Wardner.

[22]During this period of time Orchard had his picture taken in a Salt Lake City photography gallery with two friends. A mustached Orchard was standing by a card table with a cigar in his mouth, a pistol in his right hand, his left hand clutching money. On the table is a "dead man" obviously done in by Orchard because of a card game. The third man is leaning over the "dead man." Orchard claimed the photograph was simply a gag shot; others were not quite so sure.

[23]Doolan, a slim individual, five foot eight inches, was nearly thirty at the time but much younger looking than his age indicated. His face was sharp and thin. Orchard impressed him as a man of the world and one who knew his way around the mining circles.

[24]When Bill Davis left the Coeur d'Alenes, he went to the Mother Lode country of California where he used the name of W. S. Jones. Davis then went to Bisbee, Arizona, but he found he was on the mine owners' blacklist. He arrived in Cripple Creek in July, 1901, and went to work in the Shurtloff Mine on Bull Hill. He joined the Altman Miners' Union, and in 1902 became its president. Easterly was born in Illinois on a farm, but he left for Colorado to work in the mines. He arrived in Cripple Creek in April, 1896, and joined the Altman Union where he served in practically all the offices in the local.

[25]William Easterly was a man dedicated to the belief that violence could best influence labor disputes, and he participated in such activities. Enlisting the aid of two hoodlums named Steve Adams and Arthur Baston, he severely beat the non-union foreman of the Vindicator Mine No. 2, Andy Stark. A little later, Easterly, along with Adams and Sherman Parker, caught Jack Robinson in the Findley shaft and almost beat him to death. Easterly and Adams, along with four or five other men, went over to the Economic Mill and attacked the mill superintendent and another employee. Sherman Parker, also a power in the Altman union, along with Easterly, detailed Ed Minster and Adams to kill the manager of the El Paso Mine at Cripple Creek. The original plan called for the two hoodlums to shoot the mine manager through the windows of the train as he traveled, but the whole affair was abandoned before it was carried out.

[26]The big mines on Bull Hill recognized all unions, and they in turn controlled

and influenced other union locals in the district. One way the Altman union directed affairs through the years even though it was not the largest union was through the use of the "Timber Gang." A collection of tough hoodlums and punks, the roster of the "Timber Gang" at one time or another included Steve Adams, "Slim" Campbell, H. H. McKinney, Billy Gaffney, Ed Minster, Arthur Baston, Sherman Parker, and W. F. Easterly. They were brutal men, firm believers in the use of terrorist tactics in dealing with union problems, and they never hesitated to put their principles into action at the slightest provocation.

27Orchard had roomed with another miner, John Thompson, at Mrs. Neville's boarding house. Needing money, he had broken into Thompson's trunk and had stolen a gold watch, a pearl handled revolver, and forty pounds of high grade ore. He sold the ore to an assayer and pawned the other items at Victor. Later, Orchard wrote Thompson a letter enclosing the pawn ticket, no doubt a token gesture of atonement.

28Harry Orchard was married to Ida Toney at South Altman on June 7, 1903, by L. H. Weedon, Justice of the Peace.

29The political strength of the Federation is best indicated by an examination of some of the civil officers in the mining district in 1903. Among Teller county officials, the sheriff, H. M. Robertson, several of his deputies, the coroner, county clerk, and assessor, were all affiliated. Cripple Creek had several sympathetic officials; in Victor, the day marshal, night marshal, street commissioner, fire chief, jailer, three aldermen and four policemen were union members. In Goldfield, the police magistrate, day and night marshal, and all aldermen were Federation members. The marshal of Independence and the Altman justice of the peace were affiliated. In Anaconda, the marshal and three aldermen were members.

30On September 3, 1903, Thomas Stewart, a red-bearded, stubborn Scotchman, by trade a non-union carpenter, paper hanger and painter, obtained a job building a fence around the Golden Cycle Mine. The next day on the way to work, Stewart was confronted by a group of Federation men. Ed Minster, spokesman, told him, "If I went to work I'd have to stand the consequences." Unafraid, Stewart continued. Eight o'clock that same night, Minster, Adams, Parker, "Slim" Campbell and about a dozen other Federation men dragged him out of the house at gun point, brutally injured him with fists and clubs, shot him in the back, and left him for dead. Taken to the hospital, he recovered, but he was crippled for life. Several times Stewart appealed to the district attorney concerning the incident, but nothing happened. Ed Minster was arrested in connection with the assault, but was released by the sheriff who contended that he did not want "to offend a whole 'raft' of people."

31During the Haywood trial Orchard testified regarding his abortive attempt to blow up the Vindicator Mine. Attorney Richardson questioned the witness, Harry Orchard, on behalf of the defense:

Q. How many (men) did you presume you were going to kill that time?

A. I believed it would kill everyone in the mine.

Q. How many men were there working in the mine at that time?

A. I don't know how many there was there, probably forty on a shift, maybe more.

Q. And you proposed to murder fifty or more men working on a shift for two hundred dollars?

A. I proposed to touch off the carload of powder.

Q. Didn't you care how many you killed?

A. No sir, I didn't.

[32]William Aikman joined the Western Federation of Miners in 1893. He lived in Independence as a member of the Altman union. Aikman met Orchard in Neville's saloon, and the two worked together on a contract in the Deadwood claim. Both quit after two shifts because the rock was too hard. Aikman had also worked in the Vindicator and was familiar with its underground workings.

[33]From Saturday, November 21 to Tuesday, the 24th, eighteen men were arrested in connection with the explosion and were thrown in the "Bull Pen." Among these men were Sherman Parker, William Easterly, Charles G. Kennison, W. F. Davis, and Steve Adams. Information, based on the sworn allegations of the manager of the Vindicator mine, was filed on December 3; it charged Steve Adams, W. F. Davis, Sherman Parker, and Charles G. Kennison with the murder of Beck and McCormick. The cases were later *nolle prossed* by the assistant District Attorney.

[34]After an investigation, Sheriff H. M. Robertson of Teller county concluded:
"There is no doubt but that a machine has been set consisting of a box of giant powder, caps and revolver, but whether the explosion was caused by the dropping of the cage or by a time clock we are not able to decide. However, we could not find any of the parts of a clock there. The explosion must have been something terrible, for it blew the woodwork of the station to atoms."

The most prominent theory was that a man from a work force of forty men on the eighth level came up to the sixth level, fixed the instrument of death in place, and then returned to the eighth level where he resumed work. The Coroners Jury's verdict, however, stated, "From the examination made at the mine and the evidence introduced, this jury is unable to determine the exact cause of said explosion."

The official federation explanation of the explosion was that, "It is evident that McCormick and Beck planned to bring off an explosion, as it was currently reported that the state militia was about to be ordered home, and the mine owners association was against this removal. McCormick and Beck, in planning this infernal machine, made some mistake, which resulted in their death."

[35]Thomas C. Foster was a charter member of the Burke, Idaho, union and had worked in the mining camps of Burke, Butte, Bisbee and Cripple Creek.

[36]After Pettibone was released from prison for his participation in the 1892 Coeur d'Alene riots, he migrated to Denver where he opened a novelty furnishing store, selling wringers, clocks, lamps, rugs, curtains, silverware, carpet sweepers and Bibles. His store was located at 1725 Stout Street, Denver.

[37]One of the first uses of Greek-fire was in the defense of Constantinople by the Byzantine Emperors against the invading Saracens.

[38]After Harry Orchard made his confession, the deadly ingredients were found buried where he had indicated they were. They were taken to the office of the Cripple Creek Mine Owners Association and placed in an old safe. Eventually the chemicals caught fire and burned.

[39]This argument was also used by the defense in the subsequent trial of Haywood, but this time it was argued that the witness, Harry Orchard, was bribed and frightened into making false statements.

[40]Steve Adams and Art Baston were detailed to kill Collins. Soon after the killing, Adams "showed up" in the Cripple Creek mining district. Sheriff Rutan of Telluride made the investigation of Collins' death, and along with Bulkley Wells, was instrumental in getting the state militia into the mining district in November, 1903. Martial law was declared in January, 1904.

[41]Also part of the poster was a photograph of a man handcuffed to a telephone pole. On March 2, five men were convicted of vagrancy and put to work in the streets by order of the sheriff. One of them, Harry Maki, a union miner, refused to work, so a deputy sheriff handcuffed and tied him to a telephone pole. Maki was punished in this manner from 11 a.m. until 12:45 p.m. During this time, he was photographed.

[42]John Mitchell was president of the United Mine Workers of America from 1898 to 1908. He became famous during the Pennsylvania coal strike of 1902, when he organized 150,000 immigrant miners and won better wages and shorter hours for them.

[43]The Belmont Hotel was operated by a Mr. and Mrs. Guerrin and was referred to in the language of the day as an assignation house. There was some speculation as to whether the pair was actually married. Orchard was inclined to think they were not. He frequently took Mrs. Guerrin out, and Pettibone was also intimate with her. She carried insurance on the Belmont and asked Pettibone to burn it down. He refused to do it by stating that he would never trust a woman of that kind. Pettibone seems to have met Mrs. Guerrin in Montana and had the impression she was connected with a train robbery in that state. Mr. Guerrin claimed to be a newspaper man and left the hotel business pretty much to his wife. Pettibone did his best to keep the place filled with Federation members.

[44]Witnesses stated they saw three men about ten o'clock on the night of the murder in the vicinity of the Clary saloon, and one of them was carrying a shotgun. The incident was not reported until detectives questioned them.

[45]Orchard burned down Neville's saloon, situated near the Independence depot, a short time before he and Adams destroyed the station. On the advice of his good friend, Harry Orchard, Neville had insured his place of business. One evening near midnight Orchard scattered "Pettibone's Dope" on the upper floor, closed the doors, and left the building. It proved to be much easier work than the arson job he had done on his own cheese factory. Neville paid him a hundred dollars for his work.

[46]Attorney Edmund Richardson asked Orchard about the Independence Depot explosion on the witness stand in Boise, Idaho in 1907:

Q. You expected to kill 50 or 60 men that night?
A. I did not know how many we would kill.
Q. The more the merrier?
A. I did not think about that.
Q. And you did not care?
A. I don't know as I did.

[47]James J. Wallace, attorney of Cripple Creek, an officer with the Colorado militia who served during the strike, swore he saw K.C. Sterling, the mine owner's detective, fire the first shot in the Victor riot that day.

[48]Orchard was introduced to Arthur Baston by Pettibone in the latter's store in July, 1904. The two murderers got acquainted by discussing the various crimes they had committed.

[49]Haywood was a personal friend of Mayberry; they had worked together in the Nevada mines.

[50]On September 10, a telegram was sent to H. Green in c/o Peter L. Hoff, 211 Taylor Street, S.F., Calif., payable for the amount of $97.50. It was signed J. Wolff, 1725 Stout St. Another telegram on October 15, for the amount of $48.00 was sent to H. Green, c/o Peter L. Hoff. It was sent under the name of P. Bowen.

Agency out of Spokane became part of the state investigative forces, and eventually the Pinkerton Detective Agency was involved.

[63]An official report of the search of room 19, Saratoga Hotel conducted on January 2, revealed the following:

> We found small particles of a substance on the floor supposed to be gun cotton or smoke-less powder. Also found a copper band and a small wire, which corresponded identically with those found in Hogan's effect at the jail, also pieces of copper or brass metal, identically same as those found in the wood, the trees and other buildings in the vicinity of the explosion at the gate of Ex-Governor Steunenberg. In the vessel of the room we discovered evidences of where plaster of paris had been mixed, and probably used in construction of the bomb.

[64]Moyer later admitted that at the time the Federation arranged to employ a lawyer for Orchard, there had been no preliminary investigation to determine the man's guilt or innocence or to learn if Jack Simpkins, an official of that organization, had been in Caldwell with him.

[65]When John Nugent was finally informed of Haywood's request to get ready to defend Harry Orchard, he replied, "I don't see how the Western Federation of Miners is involved." Nugent was adament in wanting an explanation.

[66]Two days after the identification was made, deputy District Attorney, C.A. Gillette of Cripple Creek filed information in the district court charging Harry Orchard with the murder of J.W. Hartsell, one of the men killed in the Independence Depot explosion.

[67]When Jack Simpkins was in Denver he was given two hundred and thirteen dollars. Simpkins took part of it and had the remaining hundred sent to him which he intended on giving to Orchard. This is the hundred dollars that was given to Miller.

[68]Hutchinson, a mining man, and somewhat of an authority on explosives, considered Orchard, "One of the best posted men in the United States on explosives."

[69]The next day, Jack Simpkins was identified from a photograph by a young College of Idaho student as the man he had seen in Caldwell in the company of Orchard and in the vicinity of the Steunenberg home prior to the murder.

[70]Miller was paid fifteen hundred dollars on account, when he came to Denver on January 30, 1906.

[71]Jack Simpkins left Spokane on January 15, 1906. Governor Gooding had given his consent for the arrest, which was to have taken place on the very day he mysteriously disappeared. Miller indicated he might have gone to his timber claim on the St. Joe River.

[72]Wells took official leave from his state duties and acted in the capacity of a private citizen to accomplish the mission.

[73]Adams registered at the Tupper House in Pocatello on September 24, 1903, using the name J. Ward. The hotel was located on the east side of town, about a block from the train depot.

[51]L. McCleary, assistant city chemist, confirmed the analysis, stating: "There was so much strychnine in the milk I could see it on the raw milk."

[52]John C. Sullivan must have been under a lot of pressure to write the letter of recommendation because he was not known to like that kind of letter.

[53]Pettibone had moved his place of business to a more desirable location on Court Place just off Seventeenth Street near the luxurious Brown Palace Hotel.

[54]When Max Malich's saloon did burn down at 11:47 on that June day in 1905, he and his family were conveniently hundreds of miles away in Melrose county.

[55]On his return to Denver, Orchard checked into the Virginia Hotel on Stout Street, between Eighteenth and Nineteenth Streets. Registering as Thomas Hogan, he seemed to enjoy the nice, quiet atmosphere of the place which was operated by two ladies named Huster and Horton. The man, when he desired, could make himself as much at home in this environment as in the more bawdy saloons of the city. Perhaps this ability helped the mass murderer remain undetected for as long as he did.

Orchard's social life also picked up with his return to the city. He renewed his acquaintance with Lottie Day, who knew him as Dempsey, and with a Jennie Wilson who worked in the dress-making department of the Daniel & Fisher Stores Company. His relationship with both women seems to have been strictly on a companionship level.

[56]Haywood wanted Moyer killed after he was released from the Telluride jail. The reason was that if he had not already talked, he would surely do so. The murder could conveniently be blamed on the Mine Owners Association. Adams was willing to do the killing, but Orchard objected and talked him out of it.

[57]Steunenberg was more obnoxious to the Federation leaders than the ordinary man. Most hatred would have died with the years, but he had gotten into office with union votes, and they felt that he should stand with them, even though illegal acts were comitted.

[58]The register of the Bi-Metallic Hotel showed that Harry Orchard registered on September 12, 1905.

[59]Paulsen was aware of the attempted kidnapping plot before Orchard's confession was made public. Mr. Robertson of the Spokane law firm of Robertson, Miller and Rosenhaupt had told him of the entire affair. Robert was the legal representative of the Western Federation of Miners and a close friend of Jack Simpkins. Simpkins, in telling the story to him, claimed to have stopped the attempt.

[60]Esmeralda County, Nevada, records show John J. Neville, age 52, died October 27, 1905, of carcenoma of the rectum.

[61]This was the second bomb constructed specifically for Steunenberg's death. The attachment of the alarm clock indicated his preoccupation with the device, but actually he had two means of detonating the bomb — the alarm clock and a pull release. Circumstances indicated the use of the latter method.

[62]The work of conducting the preliminary investigation of the crime fell legally to J.C. Nichols of Canyon County. He was assisted by Sheriff D.H. Moseley of Ada County, Sheriff Angus Sutherland of Shoshone County, Sheriff Harvey K. Brown of Baker County, Oregon, and literally dozens of deputies most of whom volunteered their services. A few days later, W.S. Swain of the Thiel Detective

[74]In January, 1907, McParland, who was in Denver at that time, received an urgent telephone call from the secretary of the Mine Owners Association. It seemed that the entire contents of the safe were on fire. The detective advised them to remove the safe from the building and gave them the combination to open it. By that time the heat from the safe had set the floor on fire and only the efforts of the fire department brought it under control by moving the metal box outside. It appeared that some of the ingredients somehow had spilled out of their containers and had mixed with other chemicals, causing the fire.

[75]At this point Darrow was unaware of the part Adams had played in the North Idaho killings, and he was shaken up considerably when he learned about it.

[76]According to Adams' confession, he went to the Coeur d'Alene at the direction of Haywood and Pettibone to assist Jack Simpkins settle his timber claims problem. Once that was completed, he was to "get" Steunenberg. Simpkins offered to help kill the ex-governor, but before the attempt could be carried out, Adams went back to Colorado to get blood money promised him and never returned to finish the job.

[77]Newt Glover hauled supplies up and down the St. Joe River and was described as a "regular river rat." He was also a fine pistol and rifle shot. Simpkins, in Adams' opinion, was a poor shot and was almost deaf, so that a man could get within ten feet of him without being heard. Simpkins had a squint in both eyes as a result of a bomb that had exploded while he was attempting assassination. For his part in the killings, Adams received two hundred dollars from Pettibone before leaving Denver, and later he received one hundred from Mason and twenty dollars from Simpkins.

[78]A notice to claim jumpers was posted on trees; it contained a warning and showed a skull and cross-bones, dated August 10, 1904. Such notice was on a tree near Simpkins' cabin.

[79]The popularity of the prisoners was high in many segments of society as evidenced by a political contest. The previous year Haywood had been the Socialist candidate for governor of Colorado and had polled an amazing sixteen thousand votes, eight times as many as any other Socialist candidate. At a parade in Chicago, thousands of little girls, among them adult men and women, expressed sympathy for Haywood's daughter, Henrietta. They carried red banners which read, "Will they hang my papa?"

[80]The killer received a great deal of liberty at the penitentiary — special treatment reserved for the elite and important, even if he was a member of the criminal element. Orchard's meals came from Warden Whitney's kitchen, and sometimes he ate with the warden. In fact the prisoner wrote Hawley a letter in which he used his influence to get Warden Whitney more money when Whitney had threatened to quit. Orchard had power since he was the key to the State's case, and he knew it. Governor Gooding called him Harry; a new suit of clothes was made for him by convict labor. His hat and his linens were purchased by Whitney, and he had worn an overcoat that belonged to the warden. The murderer was an important man.

[81]Several of the jurors expressed their belief that none of them thought Haywood innocent. They voted for acquittal because the court instructions had called for such action.

[82]For some unexplained reason, the prosecution never really pursued the fact that the revolver found on Orchard at the time of his arrest had been traced to a Denver hardware store where it had been purchased by Pettibone.

[83]The last attempt at a pardon was made in 1930. His application for clemency was based on, "a long splendid record and complete reformation." Orchard left proof of his assertion, "To the officers of the prison some of whom have known me for twenty years." It was not good enough, and the State Board of Pardons issued the final verdict spelled out in cold black type. "The application for pardon of Harry Orchard, prisoner no. 1406, was denied."

SELECTED BIBLIOGRAPHY

Papers, Reports, Court Cases, Official Documents

Idaho. "Confessions of Harry Orchard and Steve Adams." James H. Hawley Papers. Idaho Historical Society, Boise.

Idaho. "Confession of Harry Orchard." Idaho Historical Society. Boise. A typescript draft of Orchard's statement to James McParland. This draft is not the original confession of January, 1906, but was an expansion of that document. It is earlier than the version published after the Haywood trial.

Idaho. "The Closing Argument of W. E. Borah." Haywood Trial. *The Statesman* Shop, Boise. No Date.

Idaho. *William E. Borah Papers,* Idaho Historical Society. Boise.

Idaho. The Closing Argument of W. E. Borah for the Prosecution. In the Great Coeur d'Alene Riot-Murder Trial. Delivered July 27, 1899, Wallace, Idaho.

Idaho. Idaho Historical Society. Boise. *James H. Hawley Papers.*

Idaho. Proceedings of the trial of William Dudley Haywood on an indictment of first degree murder, in the case of State v. Haywood, June-July, 1907. Transcript, Idaho Historical Society, Boise.
In re Boyle, 47 Pacific 706.
In re Corcoran, 59 Pacific 19.
Pettibone v. Nichols, 203 U.S. 111.
Pinkerton Detective Agency Reports relating to the Western Federation of Miners, 1906-1907. From the Hawley and Borah papers. Included are several Thiel Detective Agency reports. Idaho Historical Society, Boise.

Idaho. Idaho Historical Society, Boise. "Report of the proceedings of the special advisory board appointed by Governor James H. Peabody to investigate labor difficulties existing in the State of Colorado and more particularly at Colorado City." Denver, 1903.

U.S. Congress. House. "Coeur d'Alene Labor Troubles." H.R. 1999, 1st Session, 56th Congress.

U.S. Congress. Senate. *Report of Brigadier General H. C. Merriam on Miners' Riots in the State of Idaho,* Senate Documents 24, 142, 56th Congress, 1st Session.

U.S. Congress. Senate. Edward Boyce, President, Western Federation of Miners: Crime of the Century. Document 25. 56th Congress, 1st Session, Senate Document 25.

U.S. Congress. Senate. Reply by John L. Kennedy, Member of the Industrial Commission, which visited Idaho. (Reply to Senate Document 25). Senate Document 42, 56th Congress, 1st Session.

Books

Adamic, Louis. *Dynamite.* New York: The Viking Press, 1931.

Asbury, Herbert. *The Barbary Coast.* New York: Garden City Publishing Company, Inc., 1933.

Bankson, Russell A. and Lester S. Harrison. *Beneath These Mountains.* New York: Vantage Press, 1966.

Barrymore, Ethel. *Memories, An Autobiography of Ethel Barrymore.* New York: Harper and Brothers, 1955.

Beal, Merrill D. and Merle W. Wells. *History of Idaho.* 3 vols. New York: Lewis Historical Publishing Co., 1959.

Brissenden, Paul F. The I.W.W., A Study of American Syndicalism. New York: Columbia University, 1920.

Broehl, Wayne G. *The Molly Maguires.* Cambridge: Harvard University Press, 1964.

Brown, Robert L. *An Empire of Silver.* Caldwell, Idaho: Caxton Printers, Ltd., 1968.

Conlin, Joseph R. *Big Bill Haywood and the Radical Union Movement.* New York: Syracuse University Press, 1969.

Darrow, Clarence. *The Story of My Life.* New York: Charles Scribner's Sons, 1934.

Fisher, Vardis and Opal Laurel Holmes. *Gold Rushes and Mining Camps of the Early American West.* Caldwell, Idaho: The Caxton Printers, Ltd., 1968.

French, Hiram T. *History of Idaho.* Chicago: Lewis Publishing Co., 1914.

Fahey, John. *The Ballyhoo Bonanza.* Seattle: University of Washington Press, 1971.

Greenough, W. Earl. *Coeur d'Alene Mining Region,* 1849-1946. Mullan, Idaho: W. Earl Greenough, 1947.

Greever, William S. *The Bonanza West.* Norman: University of Oklahoma Press, 1963.

Grover, David H. *Debaters and Dynamiters.* Corvallis: Oregon State University Press, 1964.

Hammond, John Hays. *Autobiography.* New York: Farrar & Rinehart, Inc., 1935.

Hawley, James H. *History of Idaho.* Chicago: S. J. Clarke Publishing Co., 1920.

Haywood, William D. *Bill Haywood's Book.* New York: International Publishers, 1929.

Holbrook, Stewart H. *The Rocky Mountain Revolution.* New York: Henry Holt & Co., 1956.

Horan, James D., and Swiggett, Howard. *The Pinkerton Story.* New York: G. P. Putnam's Sons, 1951.

Horan, James D. the Pinkertons: *The Detective Dynasty That Made History.* New York: Crown Publishers, Inc., 1969.

Hutton, May Arkwright. *The Coeur d'Alenes, or a Tale of the Modern Inquisition in Idaho.* Denver: The APP Engraving and Printing Co., 1900.

Jensen, Vernon H. *Heritage of Conflict*. Ithaca, New York: Cornell University Press, 1950.

Johnson, Claudius. *Borah of Idaho*. New York: Longmans, Green and Co., 1936.

Kornbluh, Joyce L., ed. *Rebel Voices*. Ann Arbor: The University of Michigan Press, 1964.

Langdon, Emma F. *the Cripple Creek Strike*, 1904. Denver: Great Western Publishing Co., 1904.

Lavender, David. *The Rockies*. New York: Harper & Row, 1968.

Lee, Mabel Barbee. *Cripple Creek Days*. New York: Doubleday and Co., Inc., 1958.

Lewis, Arthur H. *Lament for the Molly Maguires*. New York: Harcourt, Brace & World, Inc., 1964.

Livingston-Little, D. E. *An Economic History of North Idaho*, 1800-1900. Los Angeles: Journal of the West, 1965.

Lord, Walter. *The Good Years*. New York: Harper and Brothers, 1960.

MacLane, John F. *A Sagebrush Lawyer*. New York: Pandrick Press, 1953.

Madison, Charles A. *American Labor Leaders*. New York: Harper & Brothers, 1950.

Magnuson, Richard G. *Coeur d'Alene Diary*. Portland: Metropolitan Press, 1968.

McKenna, Martin C. *Borah*. Ann Arbor, Michigan: University of Michigan Press, 1961.

Orchard, Harry. *The Confessions and Autobiography of Harry Orchard*. New York: McClure Co., 1907.

Orchard, Harry and LeRoy E. Froom. *The Man God Made Again*. Nashville: Southern Publishing Co., 1952.

Ravitz, Abe C. and James N. Primm, ed. *The Haywood Case*. San Francisco: Chandler Publishing Co., 1960.

Reik, Theodor. *The Compulsion to Confess*. New York: Farrar, Straus and Cudahy, 1959.

Renshaw, Patrick. *The Wobbies*. New York: Doubleday and Co., Inc., 1967.

Richard, T. A. *A History of American Mining*. New York: McGraw-Hill Book Co., 1932,

Richard, T. A. *The Bunker Hill Enterprise*. San Francisco: Mining and Scientific Press, 1921.

Siringo, Charles A. *A Cowboy Detective*. New York: J. S. Ogilvie Publishing Co., 1912.

Siringo, Charles A. *Riata and Spurs*. Boston: Houghton Mifflin Co., 1927.

Siringo, Charles A. *Two Evil Isms: Pinkertonism and Anarchism*. Chicago: Charles A. Siringo, Publisher, 1916.

Sloane, Howard N., and Lucille L. *A Pictorial History of American Mining*. New York: Crown Publishers, Inc., 1970.

Smith, Duane A. *Rocky Mountain Mining Camps*. Bloomington: Indiana University Press, 1967.

Smith, Robert Wayne. *The Coeur d'Alene Mining War of 1892.* Corvallis: Oregon State University Press, 1961.

Sprague, Marshall. *Money Mountain, The Story of Cripple Creek Gold.* Boston: Little, Brown and Co., 1953.

Stoll, William T., as told to H. W. Whicker. *Silver Strike.* Boston: Little, Brown and Co., 1932.

Stone, Irving. *Clarence Darrow for the Defense.* Garden City, New York: Doubleday and Co., Inc., 1941.

Sullivan, Mark. *Our Times.* New York: Charles Scribner's Sons, 1926.

Thomas, Gordon and Max Morgan Witts. *The San Francisco Earthquake.* New York: Stein and Day, 1971.

Thompson, Fred. Compiler. *The I.W.W., Its First Fifty Years.* Chicago: Industrial Workers of the World, 1955.

Wardner, James F. *Jim Wardner of Wardner, Idaho.* New York: Anglo-American Publishing Co., 1900.

Waters, Frank. *Midas of the Rockies.* New York: Covici Friede, 1937.

Watkins, T. H. *Gold and Silver in the West.* Palo Alto: American West Publishing Co., 1971.

Weinberg, Arthur (ed.). *Attorney for the Damned.* New York: Simon and Shuster, 1957.

Wolle, Muriel Sibell. *Stampede to Timberline.* Boulder, Colorado: University of Colorado Press, 1949.

Newspapers

Anaconda Standard (Anaconda, Montana)

Butte Miner (Butte, Montana)

Caldwell Tribune (Caldwell, Idaho)

Coeur d'Alene Sun (Murray, Idaho)

Cripple Creek Weekly Journal

DeLamar Nugget (DeLamar, Idaho)

Denver Post

Denver Times

Evening Capital News (Boise, Idaho)

Evening Telegram (Portland, Oregon)

Idaho Daily Statesman (Boise, Idaho)

Idaho State Journal (Pocatello, Idaho)

Idaho Tribune (Wallace, Idaho)

Kansas City Times

New York Times

New York World

Oregonian (Portland, Oregon)

Owyhee Avalanche (Silver City, Idaho)

Rocky Mountain News (Denver, Colorado)
San Francisco Chronicle
San Francisco Daily Examiner
San Francisco Evening Bulletin
Spokesman-Review (Spokane, Washington)

Articles, Pamphlets, and Manuscripts

Anderson, Bryce W. "The Bomb at the Governor's Gate," *American West*, II, 1965.

Bailey, Joe. "The Dramatic Story of the Steunenberg Murder," *Spokesman-Review Magazine*, January 1, 1956.

Baker, Ray Stannard. "The Reign of Lawlessness," *McClure's Magazine*, May, 1904.

Colorado. Colorado Mine Operators' Association. "Criminal Record of the Western Federation of Miners: Coeur d'Alene to Cripple Creek, 1894-1904." Denver, 1904.

Colorado. Proceedings of the trial of Sherman Parker, Thomas Foster, and W. F. Davis, charged with two unsuccessful attempts to wreck the Florence and Cripple Creek Railroad. In the case of the People of the State of Colorado vs. Parker, Foster and Davis, 1904.

"Confession of Harry Orchard," *Current Literature*, July, 1907.

Connolly, C. P. "A Little Drama Out in Idaho," *Collier's*, December 7, 1907.

———. "Moyer-Haywood Case: The Story of the Idaho Mining Troubles," *Collier's*, May 11, 1907.

———. "The Moyer-Haywood Case: The Murder and the Arrest of Orchard," *Collier's*, May 18, 1907.

———. "The Moyer-Haywood Case: The Kidnapping," *Collier's*, May 25, 1907.

———. "The Moyer-Haywood Case: Harry Orchard and His Story," *Collier's*, June 22, 1907.

———. "The Moyer-Haywood Case: The Colorado Labor War," *Collier's*, June 29, 1907.

———. "The Moyer-Haywood Case: The Colorado Labor War," *Collier's*, July 6, 1907.

———."The Moyer-Haywood Case: The Colorado Labor War," *Collier's*, July 20, 1907.

———. "The Moyer-Haywood Case: What Has Been Brought Out in Haywood's Trial," *Collier's*, July 27, 1907.

Cowart, B. T. "James McParland and the Haywood Case," *Idaho Yesterdays*, Fall, 1972.

"Effects of the Haywood Verdict," *Current Literature*, September, 1907.

Feitz, Leland. "Cripple Creek." Denver: The Golden Bell Press, 1972.

———. "Cripple Creek Railroads." Denver: The Golden Bell Press, 1968.

Grant, Luke. "Haywood Trial: A Review," *Outlook*, August 24, 1907.

Grimmett, R. G. "The Regeneration of Harry Orchard," *The West*, October, 1969.

Hard, William. "The Western Federation of Miners," *Outlook*, May 19, 1906.

Hawley, James H. Arguments, Presented in Favor of Commutation: At the Hearing Before the Pardon Board of the State of Idaho in the Matter of the Application of Harry Orchard, (Boise, Idaho, 1922).

. "Steve Adams' Confession and the State's Case Against Bill Haywood," *Idaho Yesterdays*, Winter, 1963-1964.

"Haywood's Acquittal," *Nation*, August, 1907.

"Haywood's Trial," *Outlook*, June 22, 1907.

"History of the Case," *Current Literature*, June, 1907.

Kissane, Leedice. "Steve Adams, The Speechless Witness," Idaho Yesterdays, Fall, 1960.

Kizer, Benjamin. "May Arkwright Hutton," *Pacific Northwest Quarterly*, April, 1966.

Koelsch, Charles F. "The Haywood Case." Boise, Idaho Mining Association, 1946.

Mitchell, J. Paul. "An Investigation of Governor James H. Peabody's Role in the Strike at Cripple Creek, Colorado: 1903-1904." Unpublished thesis, University of Denver, 1960.

"Moyer-Haywood Trial," *Outlook*, April 6, 1907.

Munsterberg, Hugo. "The Third Degree," *McClure's Magazine*, October, 1907.

"Murder Charge at Labor Union's Door," *Harper's Weekly*, May 25, 1907.

Orchard, Harry. "The Autobiography of Harry Orchard." The Idaho State Penitentiary, Boise, Idaho, 1948. Mrs. Harry Marsh, owner.

. "Confessions and Autobiography of Harry Orchard," *McClure's Magazine*, July, August, September, and November, 1907.

Ostrom, Richard A. "The Role of Harry Orchard in Rocky Mountain Labor Violence 1890-1907." Unpublished report, Utah State University, 1969.

Pfeiffer, Emil W. "The Kingdom of Bull Hill," *Colorado Magazine*, Vol. XII, 1935.

Pratt, Grace R. "The Great-Hearted Huttons of the Coeur d'Alenes," *Montana Magazine, Spring, 1967.*

"Professional Assassin," *Outlook*, June 15, 1907.

Rastall, Benjamin M. "The Labor History of the Cripple Creek District, a Study in Industrial Evolution." University of Wisconsin Press, Bulletin No. 198, 1908.

Rockwell, Wilson. "Bulkley Wells," Denver Westerners' Roundup, Vol. XXII, May, 1966.

"Roosevelt and the Moyer-Haywood Trial," *Outlook*, May 4, 1907.

Scheinberg, Stephen. "Theodore Roosevelt's Undesirable Citizens," *Idaho Yesterdays*, Fall, 1960.

Stark, Charles R. "Hercules and the Day Family," *Pacific Northwesterner, Vol. IV, Winter, 1960.*

Sugg, George Graham. "Colorado Conservatives Versus Organized Labor: A Study of the James Hamilton Peabody Administration, 1903-1905." Unpublished Ph.D. dissertation, University of Colorado, 1964.

Thomas, Louis A. "The Western Federation in Colorado, 1893-1905." Unpublished thesis, Colorado State Normal School, Gunnison, Colorado, 1922.

Turner, George Kibbe. "Actors and Victims in the Tragedies," McClure's Magazine, September, 1907.

Watkins, T. H. "Requiem for the Federation," American West, III, 1966.

Western Federation Miners, comp. "Category of Crime of the Mine Operators Association. Denver, 1904.

Wood, Fremont. "The Introductory Chapter to the History of the Trials of Moyer, Haywood, and Pettibone." Caldwell, Idaho: The Caxton Priners, 1931.

PICTURE CREDITS

Idaho, Historical Society: Figs. 1, 2, 3, 4, 10, 14, 16, 17, 18, 19, 20, 21, 23, 24, 25, 26.

Library, State Historical Society of Colorado: Figs. 7, 12.

Western History Department, Denver Public Library: Figs. 6, 13.

University of Idaho Library Archives: Fig. 27.

The front cover illustration and line drawings in the text were made by Gregory Pole of Moscow, Idaho.